高等职业教育"十三五"精品规划教材（汽车制造类专业群）

汽车专业英语

主　编　颜　宇　孟　倩　张希亮　张美吉

副主编　曲金烨　叶　芳　魏春均　刘振革

中国水利水电出版社
www.waterpub.com.cn
·北京·

内 容 提 要

本书针对高等职业教育重在培养具有实践和创新能力的高等应用型人才的需求而编写。主要内容包括：汽车基础知识、汽车销售、发动机、底盘、电气设备、车身、汽车维护、汽车安全文明驾驶等。书中设立学习目标、短文翻译等学习版块，并设立了不同难易程度的看图连线、新词学习、对话练习等，以便学生课后练习。

本书可作为高职高专院校汽车运用技术专业、汽车检测与维修专业、汽车电子技术与控制专业、汽车整形技术专业、汽车定损与评估及汽车技术服务与营销专业的教学用书，也可作为成人高校、高专、职大、函大等层次的教学用书，还可以作为自学者及工程技术人员的自学用书以及普通高等院校相关专业的教学参考书。

图书在版编目（CIP）数据

汽车专业英语 / 颜宇等主编. -- 北京：中国水利水电出版社，2017.12
高等职业教育"十三五"精品规划教材. 汽车制造类专业群
ISBN 978-7-5170-6027-7

Ⅰ. ①汽… Ⅱ. ①颜… Ⅲ. ①汽车工程－英语－高等职业教育－教材 Ⅳ. ①U46

中国版本图书馆CIP数据核字(2017)第274677号

策划编辑：石永峰　责任编辑：张玉玲　封面设计：李 佳

书　　名	高等职业教育"十三五"精品规划教材（汽车制造类专业群） 汽车专业英语 QICHE ZHUANYE YINGYU
作　　者	主　编　颜　宇　孟　倩　张希亮　张美吉 副主编　曲金烨　叶　芳　魏春均　刘振革
出版发行	中国水利水电出版社 （北京市海淀区玉渊潭南路1号D座　100038） 网址：www.waterpub.com.cn E-mail: mchannel@263.net（万水） 　　　　sales@waterpub.com.cn 电话：（010）68367658（营销中心）、82562819（万水）
经　　售	全国各地新华书店和相关出版物销售网点
排　　版	北京万水电子信息有限公司
印　　刷	三河市铭浩彩色印装有限公司
规　　格	184mm×240mm　16开本　8.25印张　185千字
版　　次	2017年12月第1版　2017年12月第1次印刷
印　　数	0001—3000 册
定　　价	24.00元

凡购买我社图书，如有缺页、倒页、脱页的，本社营销中心负责调换

版权所有·侵权必究

高等职业教育"十三五"精品规划教材(汽车制造类专业群)

丛书编委会

主　任　于明进

委　员　(按姓氏笔画)

刁立福　王　磊　王林超　王国林

王宝安　叶　芳　田秋荣　冉广仁

白秀秀　刘家琛　刘照军　孙　菲

李清民　吴芷红　何全民　张玉斌

张玉斌　陈　聪　郑　磊　赵长利

赵培全　郭荣春　曾　鑫　颜　宇

潘　毅

前 言

　　随着我国汽车工业的发展突飞猛进，我国汽车产销量和保有量大大增加，汽车在人们生活中的地位也越来越高。为了满足大家学习汽车专业英语的需求，汽车专业英语这门课程应运而生。本教材完整介绍了汽车售前售后相关工作岗位的专业英语，使学习者能更多地了解汽车专业英语知识。

　　本书主要内容包括：汽车基础知识、汽车销售、发动机、底盘、电气设备、车身、汽车维护、汽车安全文明驾驶等。教材紧密结合汽车专业知识，通过有针对性的讲解，使学生既学习了专业知识，又学会了英语知识。

　　本书由山东技师学院颜宇、孟倩、张希亮、张美吉任主编，由山东技师学院曲金烨、叶芳、魏春均、刘振革任副主编，全书由颜宇统稿。

　　本书在编写过程中还参考了许多国内外出版的书籍、发表的报刊、网站等相关内容，在此对原作者、编译者表示由衷的感谢。

　　由于编者的水平有限，书中难免会存在某些差错，恳请广大读者提出宝贵的意见和建议，以便再版时修订改正。

编　者
2017 年 10 月

目 录

前言

Unit 1　Basic Knowledge of Automobiles ········· 1

Lesson 1　Automobile Types ················· 1
　　Learning Goals ······················· 1
　　Part One: Look and Match ············· 1
　　Part Two: Look and Learn ············· 2
　　Part Three: Exercises ················· 4
　　Part Four: Learn More ················ 5

Lesson 2　Makes and Logos of Automobiles ····· 7
　　Learning Goals ······················· 7
　　Part One: Look and Match ············· 7
　　Part Two: Look and Learn ············· 8
　　Part Three: Exercises ················· 9
　　Part Four: Learn More ················ 12

Lesson 3　Manufactures and Names of Automobiles ················· 13
　　Learning Goals ······················· 13
　　Part One: Look and Match ············· 13
　　Part Two: Look and Learn ············· 14
　　Part Three: Exercises ················· 18

Lesson 4　A Brief History of the Automobile ······ 20
　　Learning Goals ······················· 20
　　Part One: Look and Match ············· 20
　　Part Two: Look and Learn ············· 21
　　Part Three: Exercises ················· 22
　　Part Four: Learn More ················ 23

Unit 2　Automobile Sales ··········· 24

Lesson 1　Layout and Type of Works in 4S Store ·················· 24
　　Learning Goals ······················· 24
　　Part One: Look and Match ············· 24
　　Part Two: Look and Learn ············· 25
　　Part Three: Exercises ················· 26
　　Part Four: Learn More ················ 26

Lesson 2　Automobile Sales Process ··········· 28
　　Learning Goals ······················· 28
　　Part One: Look and Match ············· 28
　　Part Two: Look and Learn ············· 29
　　Part Three: Exercises ················· 30
　　Part Four: Learn More ················ 31

Lesson 3　Auto Financing ··················· 33
　　Learning Goals ······················· 33
　　Part One: Look and Match ············· 33
　　Part Two: Look and Learn ············· 34
　　Part Three: Exercises ················· 35
　　Part Four: Learn More ················ 36

Unit 3　The Engine ·········· 37
Lesson 1　The Types of Engine ·········· 37
　　Learning Goals ·········· 37
　　Part One: Look and Match ·········· 37
　　Part Two: Look and Learn ·········· 38
　　Part Three: Exercises ·········· 40
　　Part Four: Learn More ·········· 41
Lesson 2　The Overall Construction of
　　　　　an Engine ·········· 42
　　Learning Goals ·········· 42
　　Part One: Look and Match ·········· 42
　　Part Two: Look and Learn ·········· 44
　　Part Three: Exercises ·········· 45
　　Part Four: Learn More ·········· 47
Lesson 3　Engine Operating Principles ·········· 48
　　Learning Goals ·········· 48
　　Part One: Look and Match ·········· 48
　　Part Two: Look and Learn ·········· 49
　　Part Three: Exercises ·········· 50
Lesson 4　Exhaust of Engine ·········· 52
　　Learning Goals ·········· 52
　　Part One: Look and Match ·········· 52
　　Part Two: Look and Learn ·········· 53
　　Part Three: Exercises ·········· 54

Unit 4　The Chassis ·········· 55
Lesson 1　The Transmission ·········· 55
　　Learning Goals ·········· 55
　　Part One: Look and Match ·········· 55
　　Part Two: Look and Learn ·········· 56
　　Part Three: Exercises ·········· 57
　　Part Four: Learn More ·········· 58
Lesson 2　The Steering System ·········· 59
　　Learning Goals ·········· 59
　　Part One: Look and Match ·········· 59
　　Part Two: Look and Learn ·········· 60

　　Part Three: Exercises ·········· 61
　　Part Four: Learn More ·········· 62
Lesson 3　The Brake System ·········· 63
　　Learning Goals ·········· 63
　　Part One: Look and Match ·········· 63
　　Part Two: Look and Learn ·········· 64
　　Part Three: Exercises ·········· 65
　　Part Four: Learn More ·········· 67

Unit 5　Electrical Equipment ·········· 68
Lesson 1　Battery and Generator ·········· 68
　　Learning Goals ·········· 68
　　Part One: Look and Match ·········· 68
　　Part Two: Look and Learn ·········· 69
　　Part Three: Exercises ·········· 70
Lesson 2　Starter and Air-conditioner ·········· 72
　　Learning Goals ·········· 72
　　Part One: Look and Match ·········· 72
　　Part Two: Look and Learn ·········· 73
　　Part Three: Exercises ·········· 74
Lesson 3　Illumination, Signal Devices
　　　　　and Instruments ·········· 76
　　Learning Goals ·········· 76
　　Part One: Look and Match ·········· 76
　　Part Two: Look and Learn ·········· 77
　　Part Three: Exercises ·········· 78

Unit 6　The Car Body ·········· 80
Lesson 1　The Types of Car Body ·········· 80
　　Learning Goals ·········· 80
　　Part One: Look and Match ·········· 80
　　Part Two: Look and Learn ·········· 81
　　Part Three: Exercises ·········· 82
Lesson 2　The Dimension of Car Body ·········· 83
　　Learning Goals ·········· 83
　　Part One: Look and Match ·········· 83
　　Part Two: Look and Learn ·········· 84

  Part Three: Exercises ·················· 85
  Part Four: Learn More ··············· 86
Lesson 3 Auto Interiors and Exteriors ············ 87
  Learning Goals ························· 87
  Part One: Look and Match ·············· 87
  Part Two: Look and Learn ············· 88
  Part Three: Exercises ·················· 89
  Part Four: Learn More ················ 90
Unit 7 Car Maintenance ··············· 92
Lesson 1 Car Maintenance Tools ············· 92
  Learning Goals ························· 92
  Part One: Look and Match ·············· 92
  Part Two: Look and Learn ············· 94
  Part Three: Exercises ·················· 95
  Part Four: Learn More ················ 96
Lesson 2 Car Maintenance Measuring Tools ···· 98
  Learning Goals ························· 98
  Part one: Look and Match ·············· 98

  Part Two: Look and Learn ············· 99
  Part Three: Exercises ················· 100
  Part Four: Learn More ··············· 100
Lesson 3 Complete Maintenance ············· 102
  Learning Goals ························ 102
  Part One: Look and Match ············· 102
  Part Two: Look and Learn ············ 103
  Part Three: Exercises ················· 104
  Part Four: Learn More ··············· 105
附录 练习题答案 ······················· 107
  Unit 1 Basic Knowledge of Automobiles ···· 107
  Unit 2 Automobile Sales ················ 110
  Unit 3 The Engine ······················ 111
  Unit 4 The Chassis ····················· 114
  Unit 5 Electrical Equipment ············· 116
  Unit 6 The Car Body ···················· 118
  Unit 7 Car Maintenance ················· 120

Unit 1 Basic Knowledge of Automobiles

Lesson 1 Automobile Types

Learning Goals

Understand the different types of automobiles.
Grasp the words of common types of automobiles.

Part One: Look and Match

Look at the pictures of different types of automobiles and match them to the right names given below.

bus   convertible   jeep   truck   SUV   sedan   sweeper   pickup   ambulance   fire engine   van   motor homes

1. _____

2. _____

3. _____

4. _____

5. _____

6. _____

7. _____ 8. _____ 9. _____

10. _____ 11. _____ 12. _____

Part Two: Look and Learn

A. Text

Types of automobiles

There are various kinds of automobiles for different uses. Some are used to carry passengers, some are used to deliver goods, some are used for special purposes. Therefore, automobiles can be classified into 3 groups:

1. Passenger Vehicle

Passenger vehicles are usually refer to family cars. They are used to carry no more than 9 people including the driver. Sedan, jeep, SUV(Sport Utility Vehicle), MPV(Multipurpose Passenger Vehicle), ORV(Off-Road Vehicle), pickup and minivan are all included in this group.

2. Commercial Vehicle

Commercial vehicles are designed to carry passengers (more than 9 seats) and deliver goods. It can be subdivided into bus and truck.

3. Special-purpose Vehicle

Special-purpose vehicles refer to automobiles equipped with special equipment. Road sweeper, water sprinkler, fire engine, ambulance, police car are all special-purpose vehicles.

B. New words and Phrases

automobile 汽车
type 种类，类型
be used to 被用来……
passenger 乘客
goods 货物
purpose 目的
be classified into 被分（类）为……
include 包括，包含
jeep 吉普车
ORV (Off-Road Vehicle) 越野车
minivan 面包车
be equipped with 安装有……
road sweeper 道路清扫车
fire engine 救火车
police car 警车
MPV(Multipurpose Passenger Vehicle) 多功能车

vehicle 车辆
various 各种各样的
carry 搬运，携带
deliver 交付，递送
special 特殊的，特别的
commonly 通常地
be refer to 指的是
sedan 轿车
SUV (Sport Utility Vehicle) 运动型多用途车
pickup 轿卡
commercial 商业的，盈利的
equipment 设备，装置
water sprinkler 洒水车
ambulance 救护车

C. Translation

汽车类型

由于用途不同，汽车的种类也繁多。有的用于载客，有的用于运输货物，有的则有专门的用途。因此，汽车通常可以分为三种类型：

1. 乘用车

乘用车通常指的是家庭用车，用于承载包括司机在内的至多九名乘客。轿车、吉普车、运动型多用途车、多功能车、越野车、轿卡和面包车都属于乘用车。

2. 商用车

商用车是指用于运送乘客和货物的车辆，可以细分为公共汽车和卡车。

3. 专用车

专用车是指安装有特殊装置的汽车。道路清扫车、洒水车、消防车、救护车、警车都是专用车。

Part Three: Exercises

A. What are we? Please write down the English and Chinese names of the following vehicle pictures.

_____
_____

_____
_____

_____
_____

_____
_____

_____
_____

_____
_____

_____
_____

_____
_____

_____
_____

_____
_____

_____
_____

_____
_____

Unit 1 Basic Knowledge of Automobiles

B. Fill the table with the types of automobiles by the classification of the following groups.

Types of Automobiles

Passenger sedan	Bus	Truck	Special-purpose Vehicle

C. Game: Radish down of automobile types.

Rules: 5 students in one group, the teacher give 5 cards which bear the names of different types of automobiles. Each student choose one card, tell all the class your "name" ("I am a bus" for example), then the teacher designate one of them to start the game. The group which can keeps the longest round wins.

Part Four: Learn More

A. 汽车分类

汽车的分类方式有很多种，最常见的是课文中的按照用途分类，主要分为三大类：乘用车、商用车和专用车。

根据动力装置进行分类，有内燃机汽车（Internal Combustion Engine Automobile/ICE Vehicle）、电动汽车（BEV/Electric Vehicle）、燃气涡轮机汽车（Gas Turbine Automobile）。

按照布置方式进行分类，可以分为FF、FR、RR、4WD。如表1.1所示：

Table 1.1　汽车按照布置方式分类

FF	Front Engine, Front Wheel Drive (FWD)	发动机前置前轮驱动
FR	Front Engine, Rear Wheel Drive (RWD)	发动机前置后轮驱动
MR	Middle Engine, Rear Wheel Drive (RWD)	发动机中置后轮驱动
RR	Rear Engine, Rear Wheel Drive (RWD)	发动机后置后轮驱动
4WD	Front Engine, Four Wheel Drive	发动机前置四轮驱动

B. "轿车"名字的来历

先说英文名的由来。轿车的英文是 sedan。Sedan 一词原指欧洲贵族乘用的一种豪华马车，不仅装饰讲究，而且是封闭式的，可防风、雨和灰尘，安全也能得到保障。18 世纪传到美国后，也只有纽约、费城等少数大城市中的富人才有资格享用。

1908 年美国汽车大王福特推出 T 型车时，车由原来的敞开式变为封闭式，其舒适性、安全性都有很大提高，这在当时是个了不起的进步。福特在推销时很想突出他的伟大改进，于是灵机一动，将他的"封闭式汽车"（Closed car）称为 Sedan，让购车人有一种心理上的满足。从此，供普通百姓代步的普通汽车都被称为 Sedan。

再说轿车的中文名由来。其实我国古代也早就有"轿车"一词，是指用骡马拉的轿子。当西方汽车大量进入中国时，正是封闭式方形汽车在西方流行之时。那时汽车的形状与我国古代的"轿车"一样可以显出荣耀。于是，人们就将当时的汽车称为轿车。

Lesson 2   Makes and Logos of Automobiles

Learning Goals

Understand the common logos of automobiles.
Grasp the makes of automobiles.

Part One: Look and Match

Look at the pictures and write the brand name of what the logos represent for on the dash.

Volkswagen   Toyota   Buick   Honda   BMW   Benz   Hyundai
Citroen   Ford   Nissan   Audi   Chery

1. _____   2. _____   3. _____

4. _____   5. _____   6. _____

7. _____   8. _____   9. _____

10. _____ 11. _____ 12. _____

Part Two: Look and Learn

A. Dialogue

John: Hey! What are you up to on the weekend?

Henry: Not much, staying at home and watching movies, I guess.

John: Would you like to go to a Car Show with me? I heard that they have tons of new models this year.

Henry: That sounds good! Sure, I'll go. Where is it?

John: I'll meet you in front of the gate at 2 o'clock in the afternoon on Saturday.

Henry: Alright! I'll see you then!

(At the Car Show)

John: Wow! This is great! Look at these wonderful cars.

Henry: Yeah, I love driving here in Shenzhen. Do you do a lot of driving here?

John: No, I am student. but everyday I see BMW and Mercedes zooming on the Shennan road.

Henry: What is your favorite car?

John: Mine? I like sports cars, so I would say Porsche is my favorite.

Henry: It's getting late and I have class tonight, but I'll call you after class is finished, OK?

John: Great! I'll talk to you later. Bye.

Henry: Bye.

B. New words and Phrases

be up to 从事，忙于	show 展示
tons of 很多的，一大堆	model 车型
wonderful 极好的，精彩的	Mercedes-Benz 奔驰
zoom 急速上升，嗡嗡声	favorite 特别喜欢的，喜爱的
sports car 运动型车，赛车	Porsche 保时捷

C. Translation

汽车类型

约翰：嘿！这个周末你打算干什么？

亨利：没什么，可能呆在家里看电影。

约翰：想和我一起去车展吗？我听说今年有很多新车参展。

亨利：听起来不错！好啊，我过去。在哪？

约翰：周六下午 2 点我们在大门口见。

亨利：好！到时候见！

（在车展）

约翰：哇！太棒了！看这些好车。

亨利：是啊，我喜欢在深圳开车。你在深圳这儿经常开车吗？

约翰：不，不经常开车。但是每天我都能在深南大道上看见飞驰的宝马和奔驰车。

亨利：你最喜欢什么车？

约翰：我？我喜欢运动型车，所以保时捷是我的最爱。

亨利：天色已晚，我今晚还有课要上，下了课我给你打电话，好吗？

约翰：好！我晚些再和你聊。再见。

亨利：再见。

Part Three: Exercises

A. What are we? Please write down the brand names of the following car logos both in English and Chinese

_____   _____   _____

_____   _____   _____

_____

_____

_____

_____

_____

_____

_____

_____

_____

_____

_____

_____

Unit 1 Basic Knowledge of Automobiles

_____   _____   _____

_____   _____   _____

_____   _____   _____

B. Games: Pushing-hands of auto makes. 汽车品牌推手游戏

1．选择几名同学，分别担当不同的汽车品牌如"Benz""BMW""Audi""Buick""Porsche"等，贴上名牌。

2．各组双方队员要双脚并齐，面对面站立，距一臂之隔。

3．两人都伸出胳膊，四掌相对。整个游戏过程中，不允许接触搭档的其他部位。

4．每对搭档的任务是尽量让对方失去平衡，以移动双脚为准。未移动的一方将获胜。

5．获胜方重新两两组合对抗，直至冠军产生。

Part Four: Learn More

直译起来土得掉渣的汽车品牌

奔驰、宝马、讴歌……这些国外汽车品牌进入中国后往往都会给自己起一个响亮、文雅又洋气的中文名字，目的就是给消费者一个好的第一印象，最终能有好的销量。如果按照品牌英文名称直译成中文，可不像这些品牌生产的汽车一样高大上，甚至可以说土得掉渣。因此这些品牌打入中国市场时，在品牌名称翻译上还是下了一番功夫的。

首先说说大众。1933年希特勒在德国掌权时许诺民众，让每个德国家庭都能买得起汽车，为此他特地在沃尔夫斯堡（Wolfsburg）创建了大众汽车（Volkswagen）。Volkswagen 一词由"volks"（人民）和"wagen"（汽车）组成，直译过来就是"人民汽车"。

奔驰的全名叫梅赛德斯-奔驰（Mercedes-Benz），但这个名字实在太长了，不好念，因此我们就将其简称为奔驰。其实它在国外也有简称，只不过人们大都叫它的前半部分——梅赛德斯（Mercedes）。众所周知，Mercedes 是早期戴姆勒一个经销商女儿的名字，而这个源自西班牙文的名字在英文中有"mercy""merciful"的意思，翻译过来就是仁慈、慈悲。慈悲牌汽车，你还想买吗？

要说外国汽车品牌在国内的中文名，起得最好的应该就是BMW了，用古时的宝马来比喻如今的好车再贴切不过了。众所周知BMW是一个缩写，它的全称是Bayerische Motoren Werke，直译过来就是"巴伐利亚发动机制造厂"，听起来很像一家改革开放之前的国营工厂。

尽管早已"退役"多年，但提起汽车中的硬汉，人们还是会想到悍马汽车，这是一部脱胎于美国HMMWV军车的强悍机器，"悍马"的中文名是Hummer的音译，也和它的生猛卖相非常契合，其原名Hummer也是源自HMMWV军车的昵称"Humvee"。不过你可能有所不知，Hummer在英文中的意思是"发嗡嗡声之物"，或是"蜂鸟"。与它强悍的外表是不是有天壤之别？

法拉利应该是世界上最著名的跑车品牌了，这三个字甚至已经成为了超跑的代名词。我们都知道它的名字 Ferrari 来自于其创始人 Enzo Ferrari 的姓氏，不过你可能有所不知，外国的姓氏很多都是有意义的现成词汇，比如 Ferrari 在意大利语中就是"铁匠"的意思，而且还是抡大锤干粗活的铁匠。是不是一下子感觉不能再爱了呢！

Lesson 3　Manufactures and Names of Automobiles

Learning Goals

Understand the manufactures of automobiles.
Grasp the common names of automobiles.

Part One: Look and Match

What's the name of this car? Please write the Chinese name on the line.

Excelle_____　　Glof_____　　Sagitar_____

Tucson_____　　Polo_____　　Picasso_____

Corolla_____　　Jetta_____　　Accord_____

CRV_____   Camry_____   Tiguan_____

Tiida_____   Focus_____   Passat_____

Part Two: Look and Learn

A. Text

The Most Popular Cars in Different Countries in 2014

People in different countries also have different preferences for cars. Now we will compare the different orientation of different people by a set of data in 2014.

The bestseller in Germany: **Volkswagen Golf 7** （sold 255,000）

Ready sale reason: suitable price, advanced technology, safe and reliable.

Unit **1** Basic Knowledge of Automobiles

The bestseller in USA: **Ford F-series pickup truck** (sold 753,800)

Ready sale reason: off-road performance, intelligent configuration, cargo carry function, high tensile light material.

The bestseller in France: **Renault Clio IV** (sold 105,100)

Ready sale reason: Frenchman's preference to domestic cars, personality style.

The bestseller in Britain: **Ford Fiesta** (sold 10,200)
Ready sale reason: less fuel consumption, suitable size.

The bestseller in China: **Ford Focus** (sold 391,000)
Ready sale reason: compact type, suitable consumption and size.

B. New words and Phrases

popular 流行，受欢迎
compare 比较
a set of 一组，一套
bestseller 畅销商品
ready sale 畅销
price 价格
technology 技术
off-road performance 越野性能
configuration 配置，结构，外形
high tensile 高强度
domestic 国内的
less 较少的
consumption 消费，消耗
compact 紧凑的，简洁的

preference 偏爱，倾向
orientation 方向，定位
data 数据
Germany 德国
suitable 适当的，合适的
advanced 先进的
reliable 可靠的，可信赖的
intelligent 智能的，聪明的
function 功能
light material 轻质材料
personality 个性，品格
fuel 燃料
size 大小，尺寸

C. Translation

2014 不同国家最受欢迎车型大盘点

不同国家的人对汽车的偏好也各不相同。我们下面通过 2014 年的一组数据，来对比一下不同国家人的购车取向。

1. 德国最畅销车型：大众高尔夫 7（售出 25.5 万台）
畅销原因：价格合适、技术先进、安全可靠。
2. 美国最畅销车型：福特 F 系列皮卡（售出 75.38 万台）
畅销原因：越野性能、智能配置、载货功能、高强度轻质材料。
3. 法国最畅销车型：雷诺克里奥 IV（售出 10.51 万台）
畅销原因：法国人对国产车的钟爱、个性时尚。
4. 英国最畅销车型：福特嘉年华（售出 1.02 万台）
畅销原因：低油耗、合适的尺寸。
5. 中国最畅销车型：福特福克斯（售出 39.1 万台）
畅销原因：紧凑型、合适的排量和尺寸。

Part Three: Exercises

A. Translate the following automobile English names into Chinese.

1. Lacrosse
2. Avenue
3. Encore
4. Lavida
5. Passat
6. Skoda Octavia
7. Bora
8. City
9. Civic
10. FIT
11. Qashqai
12. Sonata
13. C-Quatre
14. Mondeo
15. Fiesta

B. Games: Pushing-hands of auto makes. 汽车品牌推手游戏

1. 选择几名同学，分别担当不同的汽车品牌如"Benz""BMW""Audi""Buick"

"Porsche"等，贴上名牌。
2．各组双方队员要双脚并齐，面对面站立，距一臂之隔。
3．两人都伸出胳膊，四掌相对。整个游戏过程中，不允许接触搭档的其他部位。
4．每对搭档的任务是尽量让对方失去平衡，以移动双脚为准。未移动的一方将获胜。
5．获胜方重新两两组合对抗，直至冠军产生。

Lesson 4　A Brief History of the Automobile

Learning Goals

Understand the history and development stages of automobile.

Part One: Look and Match

Look at the pictures of different types of automobiles and match them to the right names given below.

the first four-wheel car　　Audi 80 (1990s)　　Cadillac Seville (1970s)
the first Benz model　　　Panhard Dynamic in 1930s　　VW Golf (2010)
Roadmaster (1955)　　　　First Ford T-Model

1. _____

2. _____

3. _____

4. _____

Unit **1** Basic Knowledge of Automobiles

5. _____

6. _____

7. _____

8. _____

Part Two: Look and Learn

A. Dialogue

John: Hey Henry, is that you?

Henry: Oh hey, John? What's up?

John: Not too much. Just driving to school and see a shadow kinda like yours. So, yeah.

Henry: Wow, I notice that you get a new car. Damn, it's a vintage Mustang. That's sick.

John: Need a ride? Hop in. It's weird that we talk like this.

Henry: Sure, let me feel some *Fast and Furious*.

John: Sit tight.

B. New words and Phrases

shadow 影子
vintage 古老的，最佳的
sick 厌恶，恶心
hop in 进来，上车
furious 激烈的，兴奋的

kinda 有一点儿，有几分
mustang 福特野马
ride 骑，乘坐
fast 快速的

C. Translation

约翰：嘿，亨利，是你吗？

亨利：哦嘿，约翰，干嘛呢？

约翰：没啥，开车去学校。看到一个背影像是你，就叫你了。

亨利：哇，我才发现你换新车了。好家伙，复古的野马，帅爆了！

约翰：载你一程吧，快上车。咱俩这样车上车下说话太奇怪了。

亨利：好嘞，让我感受一把《速度与激情》。

约翰：坐稳了。

Part Three: Exercises

A. Match the pictures on the left with the names on the right.

Tianjin FAW

Infiniti

Geely

Haima

Brilliance

B. 游戏：猜猜我是什么车？

5~6 人一组，每组选择一款历史上比较有名气的汽车，并对此款车的所属公司、外观、性能等方面进行介绍，让其他组的成员抢答这款汽车的名字是什么，答对得 1 分答错不扣分。最后猜对最多的那一组获得课堂加分，猜出最少的一组接受游戏惩罚。

Part Four: Learn More

A. 汽车类型常见的缩写：

Table 1.3　汽车类型常见的缩写

CRV	City Recreation Vehicle	城市休闲车
SUV	Sport Utility Vehicle	运动型多用途车
MPV	Multi-Purpose Vehicle	多用途汽车
RV	Recreational Vehicle	休闲车
HRV	Healthy Recreational Vigorous	健康休闲活力车
SRV	Small Recreation Vehicle	小型休闲车
FRC	Formula Racing Car	方程式赛车

B. 汽车的基本构造

There are four basic components of the automobile: **engine, chassis, car body and electrical equipment.**

1. Engine

The engine is the heart of a vehicle and it changes fuel energy into mechanical power.

2. Chassis

The chassis is the frame of a vehicle and supports all of the major parts.

3. Car body

The car body usually used to protect the engine, passengers and goods.

4. Electrical equipment

The electrical equipment supplies electicity for many parts such as starting system, lighting system and so on.

Unit 2 Automobile Sales

Lesson 1 Layout and Type of Works in 4S Store

Learning Goals

Understand the baisc concept of 4S store.
Grasp the terms of different depatments and the type of works in 4S store.

Part One: Look and Match

Look at the pictures and match them to the departments of 4S store given below.

Information Service Department   Auto Sales Department
Spareparts Management Department   Maintenance & Repairing Department

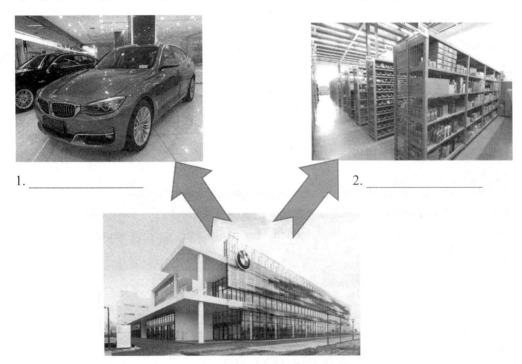

1. _____   2. _____

3. _____

4. _____

Part Two: Look and Learn

A. Text

4S Store

The 4S store is a car franchise model with the core of "four-in-one", including Sales, Spare parts, Service and Survey. It is characterised by unified appearance image, unified logo, unified management standard and only run a single brand. It is a personalized market, with consistent channels and unified cultural concepts, which can help to enhance the image of automobile brands and automobile manufacturers.

B. New words and Phrases

Information Service Department 信息服务部
Auto Sales Department 汽车销售部
Spareparts Management Department 零配件管理部
Maintenance & Repairing Department 售后维修服务部

image 形象	market 市场
spareparts 零配件	sale 整车销售
management 管理	service 售后服务
standard 标准	feedback 反馈
brand 品牌	culture 文化

C. Translation

4S 店的概念

4S 店是一种以"四位一体"为核心的汽车特许销售企业，包括整车销售（Sale）、零配件（Sparepart）、售后服务（Service）、信息反馈（Survey）等。它拥有统一的外观形象、统一的标识、统一的管理标准、只经营单一品牌的特点。它是一种个性突出的有形市场，具有一致的渠道（商品销售路线）和统一的文化理念，有助于提升汽车品牌和汽车生产企业形象。

Part Three: Exercises

A. What are we? Please write down the English names of the 4S functivns parts.

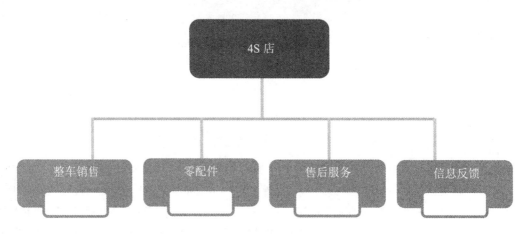

B. 请列举一下当地你所熟悉的 4S 店。（5 个）

1.
2.
3.
4.
5.

Part Four: Learn More

6S 店

现在也有 6S 店一说，除了包括整车销售（Sale）、零配件（Sparepart）、售后服务（Service）、信息反馈（Survey）以外，还包括个性化售车（Selfhold）和集拍（Sale by amount）。6S 店的兴起，得益于网络的发达，是一种利用互联网发展起来的销售模式。

个性化售车就是针对用户个人的需求来生产汽车。比如一辆越野车可以加上全景天窗，享受越野的同时又享受到敞篷车的兜风快感，也不必买一辆敞篷车而局限于其狭小的空间。当然，价格仅仅三四万块钱的车，也可以加装全景天窗。另外，更值得一提的是，如果你是爱面子的人，想拥有一辆奥迪 A6，可市场价动辄三四十万，那么这时你就可以选择个性化购车，你可以什么配置都不加，买一辆减配版的奥迪 A6，那么价格可能就只有普通价格的一半了。既然个性化售车是针对用户的个性化需求，用户就得提前下订单，利用网络的便利性，在网上轻松实现订购，订单直接传递给生产车间，生产车间按需求装配汽车。这样，一台原厂原配的个性化汽车就生产出来了。有了这一销售模式，汽车厂家按需生产，大大减少了库存及采购成本，据权威机构数据统计，这种模式可以将公司生产成本节约 30%。这样一来，个性化售车不是增加了厂家负担，而是大大降低了市场价格。

集拍，也就是集体竞拍。不难理解，这是一种直接与销量挂钩的营销模式，销量越大，价格越低。对于用户，价格上有不少优惠；对于经销商，可减少库存，减少资金积压，且可以借机增加销量。

Lesson 2　Automobile Sales Process

Learning Goals

Understand the basic concept of auto sales.
Grasp the terms of auto sales process.

Part One: Look and Match

Look at the pictures of auto sales process and match them to the right terms given below.

Reception　　　Consulting　　　Price discussion　　　Test drive
Make a deal　　Auto delivery　　After-sales service　　Automobile introduction

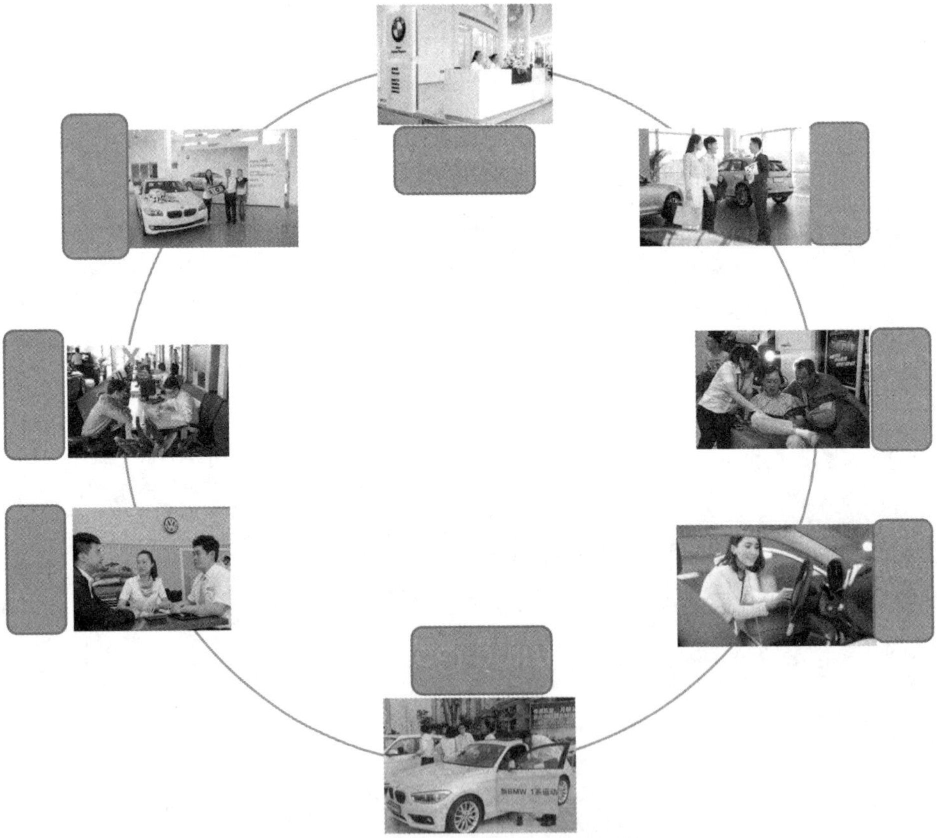

图 2.1　4S 店销售八大流程规范图

Part Two: Look and Learn

A. Dialogue

（S—Salesman，C—Customer)

S: Good morning, sir. Welcome to our 4S store.

C: Good morning.

S: I am Nick, the salesman of the store. May I know your name?

C: I'm Taylor.

S: Well, Mr. Taylor, is this your first time to visit our showroom?

C: Yes.

S: Then can I show you around? Do you buy a car for business or for your family?

C: For my family.

S: What type of car do you have in mind, sir?

C: I'm not sure. I'd like to travel with my family on weekends, so can you recommend a car for me?

S: Sure. This way, please. Look, does this SUV suit to you? It has off-road performance, large space, and the back seat can be turned into a bed.

C: Looks good. What colors does it come in?

S: We have this new model in red, white, black, silver and silver gray. These are the standard colors. What color do you prefer?

C: White, my wife prefer white.

S: I can see that you love your family very much. Good news for you that this model offers a Selfhold service, you can customize it according to your family's needs.

C: Oh, it sounds interesting.

S: Please come to my office, I can give you a detailed introduction.

C: OK.

B. New words and Phrases

Reception 接待
Automobile introduction 车辆介绍
Price discussion 报价协商
Auto delivery 交车
salesman 销售员

Consulting 需求咨询
Test drive 试驾
Make a deal 签约成交
After-sales service 售后跟踪
customer 顾客

showroom 展厅
suit 适合
customize 定做，按客户具体要求制造
recommend 推荐，介绍
seat 座椅

C. Translation

（S—销售员，C—顾客）

销售员：早上好，先生。欢迎光临我们 4S 店。

顾客：早上好。

销售员：我叫尼克，是本店的销售员。请问您贵姓？

顾客：我是泰勒。

销售员：哦，泰勒先生，这是您第一次来我们展厅吧？

顾客：是的。

销售员：那我可以带您四处看看吗？您买车是商用还是家用？

顾客：家用。

销售员：那您打算买什么类型的车呢？

顾客：我还没想好。我喜欢周末带家人一起去旅行。你给我推荐一款吧。

销售员：好的，请这边走。看，这款 SUV 适合您吗？它具有越野性能，空间大，后排座椅可以变成床。

顾客：看上去不错。都有什么颜色？

销售员：这款车的新款有红色、白色、黑色、银白色和银灰色。这些都是标准色。您喜欢什么颜色呢？

顾客：白色，我太太喜欢白色。

销售员：看得出来您很爱您的太太。告诉您个好消息，这款车提供个性化售车服务，您可以根据您和家人的需要定制。

顾客：哦，听上去很有意思呢。

销售员：请到我办公室来吧，我给您详细介绍一下。

顾客：好的。

Part Three: Exercises

A. 通过阅读以下对话，回答问题。

Dialogue:

A: Have you decided to buy that model?

B: Yes.

A: Right, Mr. Max. Come to my office please.

B: OK. Do you have a sales drive or a discount for this model?

A: I'm sorry to tell you that we have no discount and the sales drive ended yesterday.

B: Can't you postpone it for onc day?

A: No. As a famous company, you know, we must keep our promises. I am sorry.

B: That's all right.

A: So, what is your budget?

B: Then what's your price?

A: It's ￥132,000.

B: But my budget is￥130,000.

A: Frankly, we can't accept your price. If we sell it at your price, I'm afraid you won't have some spareparts. How about ￥131,500?

B: It's still more than I expected. Can't it be lower?

A: OK, ￥131,000. I can't beat that.

B: All right. It's a deal.

A: Thanks. Do you pay by credit card or in cash?

B: By credit card.

Questions:

1. What is the original price of this model? 这款车原价是多少？

2. What is the price after discussion? 打完折后这款车的价格是多少？

3. Does the customer buy the car? 顾客最后买没买车？

B. Please write down the auto sales process.

Part Four: Learn More

汽车销售中的一些常用表达：

1. What does it come with standard?
   它的标配是什么呢？

2. If you order one now, we will have it for you in August.
   如果你现在订购，我们八月份可以交车。
3. A: What is the fuel consumption of the car?
   B: It is 7.0L/100km(MT) or 7.2L/100km(AT).
   A：这辆车的油耗是多少？
   B：手动挡7升/百公里，自动挡7.2升/百公里。
4. This model sells very well.
   这款车现在很畅销。
5. The car is on sale.
   这款车是特价车。
6. 2% for full payment.
   付全款可享受九八折优惠。
7. It's a deal.
   成交。

Lesson 3　Auto Financing

Learning Goals

Understand the concept of auto financing.
Grasp the terms of auto financing.

Part One: Look and Match

Look at the pictures and match them to the right terms given below.
汽车银行贷款 bank loan
汽车金融公司 Auto Financing Company
汽车融资租赁 Auto Financing Lease
汽车厂家财务公司 Auto Manufacturer finance company
信用卡购车分期 Credit Card Installment Payment for Auto
互联网时代汽车金融 Auto financing in the Internet era

Part Two: Look and Learn

A. Text

Auto Financing

When consumers need to ask for a loan when buying cars, he can apply for the auto financing company directly for the preferential payment method. You can tailor it to your own personalized needs to choose different models and different payment methods. Compared to the bank, auto financing is a new option.

B. New words and Phrases

credit card 信用卡
financing 融资
auto financing company 汽车金融公司
insurance 保险
preferential 优先的，优惠的
method 方法
personalized 个性化的

installment 分期付款
lease 租赁
bank 银行
apply for 申请，请求
payment 支付，付款
tailor to 量身定做

C. Translation

汽车金融

汽车金融是消费者在购买汽车需要贷款时，可以直接向汽车金融公司申请优惠的支付方式，可以按照自身的个性化需求来选择不同的车型和不同的支付方式。对比银行，汽车金融是一种购车新选择。

Part Three: Exercises

A. Go online and inquire what are the current auto financing companies?

1. _____
2. _____
3. _____
4. _____
5. _____

B. Ms. Yang recently wanted to buy a BMW 320, but she doesn't have enough money. After learning this section, what do you think Ms. Yang can do to solve the shortage of money to buy the car she likes?

_____
_____
_____
_____

Part Four: Learn More

Table 2.1 2016 年排名前十汽车金融公司

1	梅赛德斯-奔驰汽车金融有限公司（Mercedes-benz financial co., LTD）
2	北京现代汽车金融有限公司（Beijing Hyundai auto financial co., LTD）
3	上汽通用汽车金融有限责任公司（SAIC-GMAC）
4	大众汽车金融公司（Volkswagen financial co., LTD）
5	丰田汽车金融（中国）有限公司（Toyota motor finance（China）co., LTD）
6	瑞福德汽车金融有限公司（Fortune Auto Finance Co., Ltd）
7	福特汽车金融公司（Ford automobile financial company）
8	东风标致雪铁龙汽车金融有限公司（Dongfeng Peugeot Citroen automobile financial co., LTD）
9	菲亚特汽车金融有限责任公司（Fiat auto finance co., LTD）
10	沃尔沃汽车金融有限公司（Volvo auto finance co., LTD）

Unit 3 The Engine

Lesson 1 The Types of Engine

Learning Goals

Understand the different classification methods of automobile engine.
Grasp the basic engine types.

Part One: Look and Match

Look at the pictures of different types of engine and match them to the right names given below (some of the pictures have more than one answers).

    two-stroke engine   four-stroke engine   air-cooled engine   water-cooled engine
    single cylinder engine   multi-cylinder engine   V-type engine   flat engine
    in-line engine   supercharged engine   new energy engine

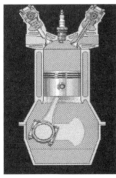
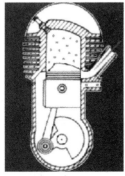

1. _____   2. _____   3. _____

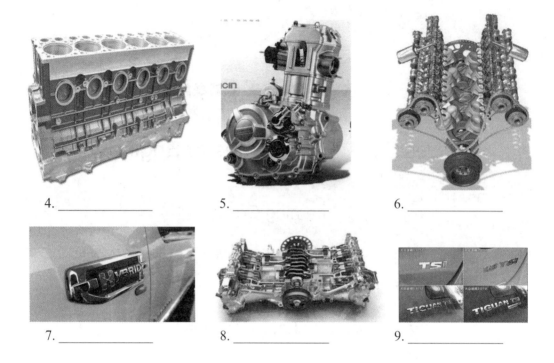

4. _____ 5. _____ 6. _____

7. _____ 8. _____ 9. _____

Part Two: Look and Learn

A. Text

The Types of Engine

According to the fuel, engines can be divided into gasoline engine, diesel engine and new energy engine, etc.

According to the stroke, engines can be divided into two-stroke engine and four-stroke engine.

According to the cooling method, engines can be divided into air-cooled engine and water-cooled engine.

According to the numbers of cylinders, engines can be divided into single cylinder engine and multi-cylinder engine. Modern automobile engines have usually four cylinders, six cylinders or eight cylinders.

According to the arrangement of cylinders, engines can be divided into single-bank engine and two-bank engine. The cylinders of two-bank engine arranged in two columns, the engine is called

V-type engine if the included angle is less than 180º, flat engine if the included angle is 180º.

According to the method of supercharging, engines can be classified into forced intake (supercharged) engine and natural aspirating (unsupercharged) engine.

B. New words and Phrases

engine 发动机	cylinder 气缸
fuel 燃料	cycle 循环
stroke 冲程，行程	cooling 冷却
air-cooled engine 风冷发动机	water-cooled engine 水冷发动机
single cylinder engine 单缸发动机	multi-cylinder engine 多缸发动机
V-type engine V 型发动机	in-line engine 直列式发动机
flat engine 对置式发动机	arrangement 排列
supercharge 增压	force 强制
natural 自然的	aspirating 吸气，吸引

C. Translation

发动机的类型

1. 按照所用燃料分类：汽油机，柴油机和新能源发动机等。

2. 按照冲程分类：二冲程发动机、四冲程发动机。

3. 按照冷却方式分类：风冷发动机、水冷发动机。

4. 按照气缸数目分类：单缸发动机和多缸发动机。现代汽车多采用四缸、六缸、八缸发动机。

5. 按照气缸排列方式分类：单列式和双列式。双列式发动机把气缸排成两列，两列夹角小于180°称为 V 型发动机，若两列夹角为 180° 则称为对置式发动机。

6. 按照是否采用增压方式分类：强制进气（增压）发动机和自然吸气（非增压）发动机。

Part Three: Exercises

A. What are we? Please write down the English and Chinese names of the following pictures.

_____   _____   _____
_____   _____   _____

_____   _____   _____
_____   _____   _____

B. Fill the table with the types of automobiles by the classification of the following groups.

The Types of Engine

Fuel	Stroke	Cooling method

The Types of Engine

Numbers of cylinders	The arrangement of cylinders	Method of supercharging

C. Game: 发动机种类萝卜蹲。

Rules: 5 students in one group, the teacher give 5 cards which bear the names of different types of engine. Each student choose one card, tell all the class your "name" ("I am a V type engine" for example), then the teacher designate one of them to start the game. The group which can keeps the longest round wins.

Part Four: Learn More

新能源汽车是指除汽油、柴油发动机之外所有使用其他能源的汽车。包括燃料电池汽车（FCEV）、混合动力汽车（HEV）、氢能源动力汽车（HFCV）、纯电动汽车（BEV）和太阳能汽车（Solar Car）等。目前中国市场上在售的新能源汽车多是混合动力汽车和纯电动汽车。分类如下表：

Table 3.1 我国新能源汽车分类

FCEV	Fuel Cell Electric Vehicle	燃料电池汽车
HEV	Hybrid Electric Vehicle	混合动力汽车
HFCV	Hydrogen Fuel Cell Vehicles	氢能源动力汽车
BEV	Battery Electric Vehicle	纯电动汽车
	Solar Car	太阳能汽车

Lesson 2   The Overall Construction of an Engine

Learning Goals

Understand the basic components of engine.
Grasp the 2 mechanisms and 5 systems of engine.

Part One: Look and Match

Look at the pictures of basic components of an engine and match them to the right names given below.

crankshaft and connecting rod mechanism   valve gear   the fuel-supply system
the lubricating system   the cooling system   the starting system
the ignition system   cylinder body   piston   connecting rod   crankshaft
flywheel   valve   oil sump   oil pump   fuel tank   starter motor   spark plug

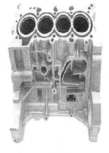

1. _____   2. _____   3. _____

4. _____   5. _____   6. _____

Unit 3 The Engine

7. _____

8. _____

9. _____

10. _____

11. _____

12. _____

13. _____

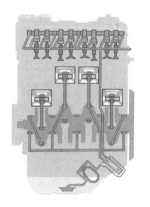
14. _____

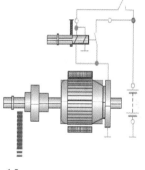
15. _____

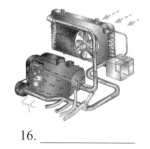
16. _____

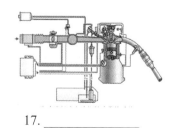
17. _____

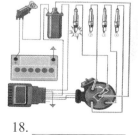
18. _____

Part Two: Look and Learn

A. Text

The Overall Construction of an Engine

The engine is composed of two mechanisms and five systems, respectively are crankshaft and connecting rod mechanism, valve gear; the fuel-supply system, the lubricating system, the cooling system, the starting system and the ignition system.

The crankshaft and connecting rod mechanism are the major moving parts to complete the operating cycle and accomplish the conversion of energy. The function of valve gear is to open and close the intake valve and exhaust valve regularly according to the working sequence and process in order to finish the gas exchange process.

The fuel-supply system makes the mixture of certain amount and concentration based on the requirements of the engine. The lubricating system lubricates the moving parts. The function of the cooling system is to release the heat of the heating parts in time to ensure that the engine works at the proper temperature. The starting system starts the engine and the ignition system works to generate spark plug and ignite combustible mixture.

B. New words and Phrases

cylinder body 气缸体
connecting rod 连杆
flywheel 飞轮
oil sump 油底壳
fuel tank 燃油箱
spark plug 火花塞
the fuel-supply system 燃料供给系统
the cooling system 冷却系统
combustible mixture 可燃混合气
crankshaft and connecting rod mechanism 曲柄连杆机构

piston 活塞
crankshaft 曲轴
valve 气门
oil pump 机油泵
starter motor 起动机
valve gear 配气机构
the lubricating system 润滑系统
the ignition system 点火系统
the starting system 起动系统

C. Translation

发动机的总体构成

发动机通常由两大机构五大系统组成，分别是：曲柄连杆机构、配气机构；燃料供给系统、润滑系统、冷却系统、起动系统和点火系统。

曲柄连杆机构是发动机实现工作循环、完成能量转换的主要运动零件。配气机构作用是根据发动机工作顺序和过程，定期开启和关闭进气门和排气门，实现换气过程。燃料供给系统作用是根据发动机的要求，配制出一定数量和浓度的混合气。润滑系统的作用是对运动零件润滑。冷却系统的作用是将受热零件的热量及时散发出去，保证发动机在适宜的温度状态下工作。起动系统的作用是起动发动机。点火系统的作用是使火花塞产生电火花，点燃可燃混合气。

Part Three: Exercises

Look at the diagrams and read the Chinese descriptions on the left. Then write down the correct engine systems in English.

1. _____

汽油机燃料供给系统的作用是根据发动机不同工况的要求，供给气缸不同浓度和数量的汽油和空气的可燃混合气。

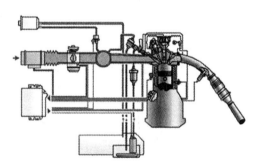

2. _____

冷却系统通过循环发动机与散热器之间的液体冷却剂使发动机得以散热。

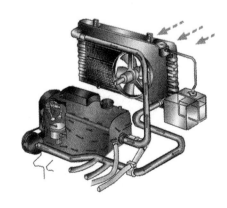

3. _____

润滑系统的作用是向发动机内的所有活动部件提供润油。润滑油使运动部件避免过度磨损。

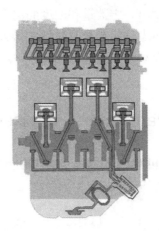

4. _____

点火系统的作用是为发动机提供高压电火花,点燃汽油发动机燃烧室内的可燃混合气。

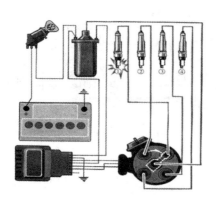

B. Match the English with Chinese. Draw lines.

1. flywheel   a. 活塞
2. spark plug   b. 连杆
3. piston   c. 曲轴
4. fuel tank   d. 飞轮
5. crankshaft   e. 机油泵
6. connecting rod   f. 油底壳
7. oil sump   g. 火花塞
8. oil pump   h. 燃油箱

Part Four: Learn More

Table 3.2 汽车部分系统中英文对照

Abbr.	English	Chinese
ECS	Engine Control System	发动机控制系统
ECM	Engine Control Module	发动机控制单元
VTEC	Variable Valve Timing and Lift Electronic Control System	可变气门正时和升程电子控制系统
RPM	Revolutions Per Minute	每分钟转速

Lesson 3 Engine Operating Principles

Learning Goals

Understand the characteristics of each stroke.
Grasp the operating principles of four-stroke engine.

Part One: Look and Match

Look at the pictures and match them to the right terms given below.

intake stroke   compression stroke   power stroke   exhaust stroke
TDC (top dead center)   BDC (bottom dead center)   combustion chamber
cylinder head   intake passage   exhaust manifold   piston ring   the four-stroke cycle

1. _____

2. _____

3. _____

4. _____

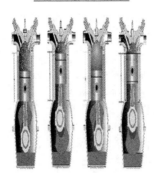

5. _____

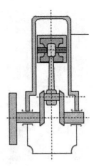

6. _____

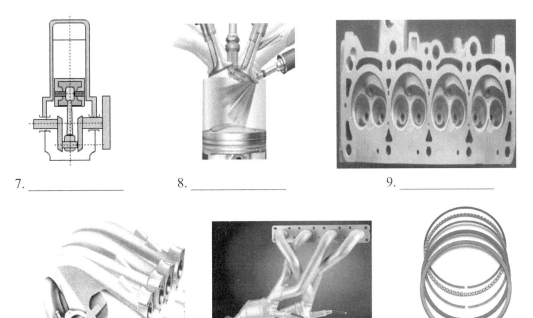

7. _____   8. _____   9. _____

10. _____   11. _____   12. _____

Part Two: Look and Learn

A. Text

Engine Operating Principles

The operation of the four-stroke engine is constantly circulating in the order of the intake stroke, compression stroke, power stroke and exhaust stroke. In each of the four-stroke cycle, the power stroke is generated, and the other three strokes are supplementary to the preparation for the power stroke.

B. New words and Phrases

operating principle 工作原理
lower dead center 下止点
combustor 燃烧室
intake stroke 进气行程
power stroke 作功行程

top dead center 上止点
cylinder head 气缸盖
the four-stroke cycle 四冲程
compression stroke 压缩行程
exhaust stroke 排气行程

intake passage 进气管
piston ring 活塞环
exhaust manifold 排气管
supplementary 补充

C. Translation

发动机工作原理

四冲程发动机的运转是按进气行程、压缩行程、作功行程和排气行程的顺序不断循环往复的。在每个工作循环的四个行程中，只有作功行程产生动力，其余三个行程都是为作功行程做准备工作的辅助行程。

Part Three: Exercises

Look at the diagrams and read the Chinese descriptions on the left. Then write down the correct engine strokes in English.

1. _____ Stroke
由于曲轴旋转，活塞从上止点向下止点运动，排气门关闭，进气门打开。

2. _____ Stroke
曲轴继续旋转，活塞从下止点向上止点运动，这时进气门和排气门都关闭，气缸内成为封闭容积，可燃混合气被压缩，当活塞运动到上止点时压缩行程结束。

3. _____ Stroke
这一行程，进气门和排气门仍然关闭，当活塞运动到接近上止点时，火花塞产生电火花点燃可燃混合气，推动活塞从上止点向下止点运动。

4. _____ Stroke
作功行程接近终了时，排气门打开，进气门仍然关闭，靠废气的压力先自行排气，活塞到达下止点再向上止点运动时，继续把废气强制排出到大气中去。活塞到达上止点后，排气门关闭，排气行程结束。

Unit 3 The Engine

B. Match the English with Chinese. Draw lines.

1. top dead center   a. 排气行程
2. lower dead center   b. 进气行程
3. the four-stroke cycle   c. 上止点
4. intake stroke   d. 四冲程
5. compression stroke   e. 压缩行程
6. power stroke   f. 作功行程
7. exhaust stroke   g. 下止点

C. Look at the picture and choose the correct technical terms.

a. exhaust valve
b. combustor
c. piston
d. intake passage
e. intake valve
f. spark plug
g. exhaust manifold
h. oil atomizer

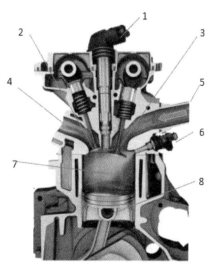

51

Lesson 4   Exhaust of Engine

Learning Goals

Understand the technology of energy conservation and emission reduction.
Grasp the main emission pollutants of the engine.

Part One: Look and Match

Look at the pictures and match them to the right terms given below.
oxygen sensor   three-way catalyst   EGR   CRDI   GDI   PVC
global warming   exhaust emission   greenhouse effect

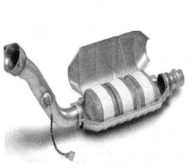

1. _____

2. _____

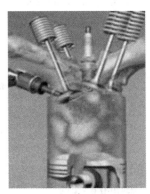

3. _____

4. _____

5. _____

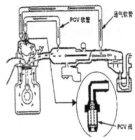

6. _____

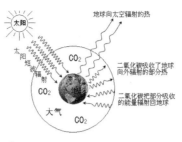

7. _____   8. _____   9. _____

Part Two: Look and Learn

A. Text

Emission pollutants

Automobile harmful emissions are tailpipe emissions and fuel system evaporative emissions. The main engine harmful emissions are CO, HC, NO_x and SO_2 and particles. CO is colorless and odorless gas which is easy to combine with hemoglobin and cause death in serious cases. NO_x is brown, has a pungent odor and produce a photochemical smog with HC.

B. New words and Phrases

CO 一氧化碳	HC 碳氢化合物
NO_x 氮氧化合物	SO_2 二氧化硫
PM 微粒	oxygen sensor 氧传感器
three-way catalyst 三元催化器	CRDI 高压共轨
GDI 缸内直喷	PVC 曲轴箱强制通风系统
EGR 废气再循环	global warming 全球变暖
exhaust emission 尾气排放	greenhouse effect 温室效应

C. Translation

发动机排放污染物

汽车有害排放物是指尾气排放物、燃油系统蒸发排放物等。发动机有害排放物主要有 CO、HC、NO_x 和 SO_2 以及微粒。CO 是无色无味气体，极易与血红蛋白结合，严重可致死亡。NO_x

为褐色、有刺激性气味，与 HC 可以形成光化学烟雾。

Part Three: Exercises

A. 写出发动机的主要排放污染物

1. _____
2. _____
3. _____
4. _____
5. _____

B. Match the English with Chinese. Draw lines.

1. PM   a. 曲轴箱强制通风系统
2. CO_2 b. 氧传感器
3. PVC   c. 微粒
4. GDI   d. 三元催化器
5. EGR   e. 缸内直喷
6. oxygen sensor   f. 二氧化碳
7. three-way catalyst   g. 废气再循环
8. CRDI   h. 高压共轨

Unit 4　The Chassis

Lesson 1　The Transmission

Learning Goals

Understand the components and classification of the transmission.
Grasp the terms of the components of the transmission.

Part One: Look and Match

Look at the pictures and match them to the right names given below.

final drive　　differential　　MT(manual transmission)　　AT(automatic transmission)
flywheel　　clutch　　friction plate　　drive shaft　　universal joint

1. _____　　2. _____　　3. _____

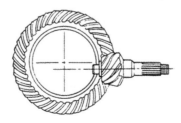

4. _____　　5. _____　　6. _____

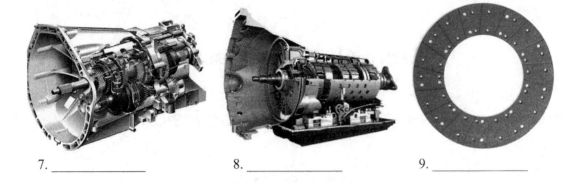

7. _____ 8. _____ 9. _____

Part Two: Look and Learn

A. Text

The Functions of the Transmission

The transmission is a set of variable speed gears used to coordinate the revolving speed of the engine and the actual speed of the wheel to develop the best performance of the engine. The transmission can produce a different ratio between the engine and the wheel when a car runs. The functions of the transmission are:

1. Change the transmission ratio.
2. Carry out the reversing motion.
3. Interrupt power transmission and idle.

B. New words and Phrases

final drive 主减速器
manual transmission 手动变速器
flywheel 飞轮
friction plate 摩擦片
universal joint 万向节
coordinate 调整，协调
ratio 比率
interrupt 打扰，中断

differential 差速器
automatic transmission 自动变速器
clutch 离合器
drive shaft 传动轴
variable speed gears 变速装置
revolving 旋转的
reverse 相反，倒退
idle 虚度，空转

C. Translation

变速器的作用

变速器是一套用于协调发动机的转速和车轮的实际行驶速度的变速装置，用于发挥发动机的最佳性能。变速器可以在汽车行驶过程中，在发动机和车轮之间产生不同的变速比。变速器的作用有：

1．改变传动比。
2．实现倒车行驶。
3．中断动力传递、实现空挡。

Part Three: Exercises

A. Look at the picture and choose the correct technical terms.

final drive   differential   manual transmission   clutch   drive shaft
universal joint   engine   halfshaft

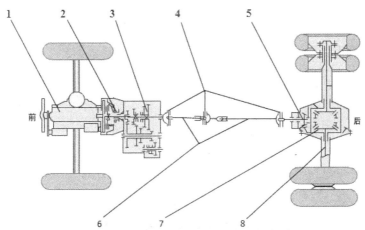

B. Complete the choice according to the conversation.

Cary : This way, Kevin. I'd like to show you the new model.

Kevin : Oh, it looks good!

Cary : Yes, it is. It has an automatic transmission and an all-wheel drive. I think it's quite a good car for young men. I bet you'll like it after a test drive.

Kevin : Very good. I love it. But its engine capacity is too large for me.

Cary : I see. You may make a decision after seeing some other models.

Kevin : Sure. Thank you.

1. Kevin is a_____.

    a. manager   b. young man   c. student

2. The car_____.

    a. has an automatic transmission

    b. doesn't have an automatic transmission

    c. has a manual transmission

3. Cary introduced_____ to Kevin.

    a. the new model   b. two models   c. several models

4. Kevin_____.

    a. and Cary drive the car   b. is a driver   c. made the decision to buy

5. Kevin_____.

    a. hasn't decided to take the new model

    b. doesn't take the new model

    c. wants to buy the car

Part Four: Learn More

Table 4.1 学习汽车上的挡位名称

挡位简称	挡位全称	汉语名称
P	Parking	驻车挡
R	Reverse	倒挡
N	Neutral	空挡
D	Driving	前进挡
S	Sport	运动挡

Lesson 2 The Steering System

Learning Goals

Understand the components and classification of the steering system.
Grasp the terms of the components of the steering system.

Part One: Look and Match

Look at the pictures and match them to the right names given below.

steering shaft   ball joint   tie rod   power steering pump   steering knuckle
steering gearbox   steering wheel

1. _____

2. _____

3. _____

4. _____

5. _____

6. _____

7. _____

Part Two: Look and Learn

A. Text

Steering System

The function of the steering system is to control the direction of the car according to the driver's wishes.

The steering system can be divided into two major categories: the mechanical steering system and the power steering system.

The mechanical steering system consists of three parts: steering control mechanism, steering gearbox and steering drive mechanism.

Power steering system is a steering system which uses both the driver's physical strength and the power form the engine as the steering energy.

B. New words and Phrases

steering shaft 转向轴
tie rod 转向横拉杆
steering knuckle 转向节
steering wheel 方向盘
mechanical 机械的
strength 力量

ball joint 转向球头
power steering pump 转向助力泵
steering gearbox 转向器
direction 方向
physical 物理的，身体的
power 力量，动力

C. Translation

汽车转向系统

汽车转向系统的作用是按照驾驶员的意愿控制汽车的行驶方向。

汽车转向系统分为两大类：机械转向系统和动力转向系统。

机械转向系统由转向操纵机构、转向器和转向传动机构三大部分组成。

动力转向系统是兼用驾驶员体力和发动机动力为转向能源的转向系统。

Part Three: Exercises

A. Read the text and complete the following sentences.

The purpose of the steering system is guiding the car to where the driver wants it to go. There are two general types of steering. One type is the manual steering system. The other type is the power steering system.

In the manual type, the driver does all the work of turning the steering wheel, the steering gearbox, wheel. In the power type, hydraulic fluid assists the operation so that the driver's effort is reduced.

The suspension system has two basic functions, to prevent the car from leaning to one side and to provide a comfortable ride for the passengers. The suspension system has two subsystems—the front suspension and the rear suspension. Modern cars use an independent front suspension.

1. The purpose of the steering system is _____.
2. There are _____ general types of steering.
3. Hydraulic fluid _____ the operation so that the driver's effort is reduced.
4. The suspension system _____ the car from leaning to one side.
5. Modern cars use an independent _____ suspension.

B. Reorder the letters with your partner.

Letters	Words
ual man	
sion sus pen	
vent pre	
dent de pen in	

Part Four: Learn More

Learn the following English abbreviations you often come across in handling in automobile materials.

Table 4.2　常用汽车词汇英文简称对照

Abbr	English	Chinese
PSCU	Power Steering Control Unit	动力转向控制单元
VSC	Vehicle Stability Control	汽车稳定控制
VIN	Vehicle Identification Number	车辆识别代码
TPMS	Tire Pressure Monitoring System	胎压监测系统
EPS	Electronic Power Steering	电子控制助力转向

Lesson 3　The Brake System

Learning Goals

Understand the components and classification of the brake system.
Grasp the terms of the components of the brake system.

Part One: Look and Match

Look at the pictures and match them to the right names given below.

brake drum　　brake pedal　　disc brake　　brake disc　　drum brake
Wheel Speed Sensor (WSS)　　ABS pump　　fluid reservoir　　parking brake
brake warning light　　backing plate　　brake caliper

1. _____

2. _____

3. _____

4. _____

5. _____

6. _____

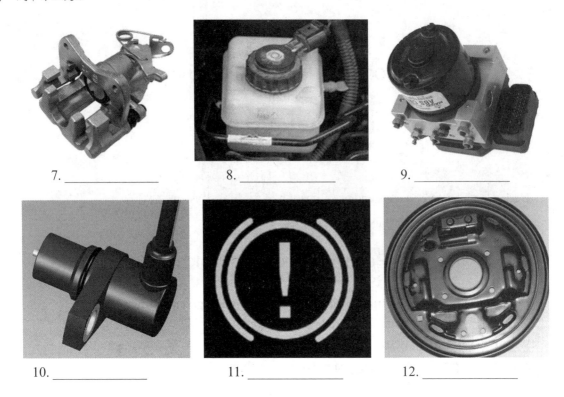

7. _____ 8. _____ 9. _____

10. _____ 11. _____ 12. _____

Part Two: Look and Learn

A. Text

Function and Classification of Brake System

When parking and braking, the braking system is installed on the vehicle to make the vehicle stop or park and ensure that the vehicle is stably parked.

Braking refers to the friction material providing external pressure power to slow down the vehicle. The friction material is on the brake drum or the brake disc.

According to the different functions, there are service brake system and parking brake system. According to the various media, there are hydraulic braking system and air braking system. The air braking system is mainly used in large capacity trucks or medium trucks.

B. New words and Phrases

brake drum 制动鼓
disc brake 盘式制动
drum brake 鼓式制动
ABS pump　ABS 泵
parking brake 驻车制动
backing plate 制动底板
stably 平稳地
external 外部的，表面的
material 材料，原料
hydraulic 液压的
media 媒介
capacity 容量

brake pedal 制动踏板
brake disc 制动盘
Wheel Speed Sensor (WSS) 轮速传感器
fluid reservoir 储液罐
brake warning light 制动警告灯
brake caliper 制动钳
friction 摩擦
slow down 减速，使……慢下来
Service Brake System 行车制动系统
Parking Brake System 驻车制动系统
various 不同的，各种各样的
medium 中等的

C. Translation

制动系统的作用与分类

停车和制动时，车辆上设置的制动系统能使车辆变为停车或驻车的状态，保证车辆稳定停放。

制动是指通过摩擦材料提供外压功率，使车辆达到降速的状态。摩擦材料在制动鼓或制动盘上。

根据功能不同，制动系统可分为行车制动系统和驻车制动系统。根据介质，有液压制动系统和空气制动系统。空气制动系统主要用于大容量货车或中型货车。

Part Three: Exercises

A. Read the text above and complete the following sentences.

1. The braking system makes sure that the vehicle can park_____.
2. The friction material affords the_____and makes the vehicle slow down.
3. The friction material is on the brake_____ or the brake_____.
4. There are the_____braking system and the_____braking system.
5. The air braking system is mainly used for_____or medium lorries.

B. Read the text again and choose the correct answer.

1. The text is about the_____.
   a. electrical system   b. transmission system   c. brake system
2. The service braking system means_____.
   a. a service station
   b. braking when the car is moving
   c. braking by hand
3. The friction material is on the_____.
   a. brake handle
   b. brake disc
   c. brake drum or brake disc
4. Parking braking means_____.
   a. handbrake
   b. power-assisted brake
   c. brake on front wheel
5. The air braking uses_____.
   a. the power of air
   b. the dielectric of air
   c. the dielectric of water

C. Look at the picture and choose the correct technical terms.

master cylinder   drum brake   disk brake   ABS pump
vacuum booster   brake pedal   brake light

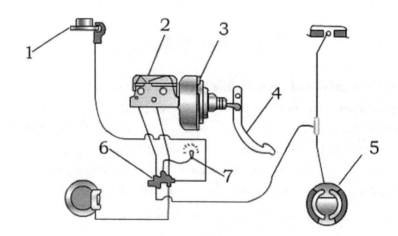

Part Four: Learn More

Learn the following English abbreviations you often come across in handling automotive materials.

Table 4.3 常用汽车词汇英文简称对照

Abbr	English	Chinese
EBD	Electric Brake-force Distribution	电子制动力分配（系统）
BMC	Brake Master Cylinder	制动主缸
ABS	Anti-lock Brake System	防抱死制动系统
BPMV	Brake Power Modulator Valve	制动压力调节器
BMC	Brake Master Cylinder	制动主缸
BPMV	Brake Power Modulator Valve	制动压力调节器

Unit 5　Electrical Equipment

Lesson 1　Battery and Generator

Learning Goals

Understand the principle of battery and generator.
Grasp the components of battery and generator.

Part One: Look and Match

Look at the pictures of the components of battery and generator and match them to the right terms given below.

plate　　separator　　electrolyte　　battery　　pole　　rotor　　stator
brush　　rectifier　　end shield　　fan　　generator

1. _____

2. _____

3. _____

4. _____

5. _____

6. _____

Unit 5 Electrical Equipment

7. _____

8. _____

9. _____

10. _____

11. _____

12. _____

Part Two: Look and Learn

A. Text

Battery and Generator

The car is equipped with two DC power sources in parallel: the battery and generator. The electrical equipments are powered by the battery when the engine starts up, and are mainly supplied by the generator when the generator works normally after the engine starts up.

Common lead-acid batteries are chiefly composed of plate, separator, electrolyte, shell, pole, etc. Common alternator consists of rotor, stator, rectifier, brush, front and rear end cover, fan and drive pulley, etc.

B. New words and Phrases

battery 蓄电池	plate 极板
separator 隔板	electrolyte 电解液
shell 外壳	pole 极桩

generator 发电机
stator 定子
brush 电刷
fan 风扇
start up 起动
consist of 由……构成
rotor 转子
rectifier 整流器
end cover 端盖
drive pulley 驱动带轮
be composed of 由……构成

C. Translation

蓄电池和发电机

汽车上装有蓄电池和发电机两个并联的直流电源。发动机起动时，由蓄电池向用电设备供电。起动后，发电机正常工作时，主要由发电机供电。

普通铅酸电池主要由极板、隔板、电解液、外壳、极桩等组成。普通交流发电机主要由转子、定子、整流器、电刷、前后端盖、风扇及驱动带轮等组成。

Part Three: Exercises

A. Match the English with Chinese. Draw lines.

1. generator   a. 蓄电池
2. rectifier   b. 极板
3. battery   c. 隔板
4. plate   d. 电解液
5. brush   e. 发电机
6. electrolyte   f. 整流器
7. separator   g. 转子
8. rotor   h. 电刷

B. Look at the picture and choose the correct technical terms.

(1)
a. positive plate
b. separator
c. shell
d. negative plate
e. pole

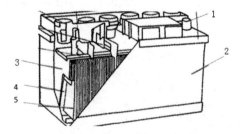

(2)

a. front end cover
b. brush
c. rectifier
d. rotor
e. stator
f. rear end cover
g. fan
h. drive pulley

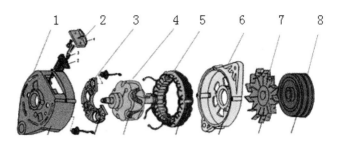

Lesson 2　Starter and Air-conditioner

Learning Goals

Understand the principles of starter and air-conditioner.
Grasp the components of starter and air-conditioner.

Part One: Look and Match

Look at the pictures of components of starter and air-conditioner and match them to the terms given below.

starter　　commutator　　armature　　condenser　　compressor　　evaporator
expansion valve　　receiver-dryer　　refrigerant

1. _____

2. _____

3. _____

4. _____

5. _____

6. _____

Unit 5 Electrical Equipment

7. _____ 8. _____ 9. _____

Part Two: Look and Learn

A. Text

Starter and Air-conditioner

The function of the starter is to start the engine. Once the engine starts, the starter stops working immediately. The starter is composed of three parts: DC series motor, drive mechanism and control mechanism.

Automobile air conditioning system mainly consists of ventilating system, heating system, refrigerating system and air cleaner. The air conditioning system consists of compressor, condenser, evaporator, receiver-dryer and blower.

B. New words and Phrases

starter motor 起动机
control mechanism 操纵机构
commutator 换向器
air conditioner 空调
heating system 采暖系统
air cleaner 空气净化装置
evaporator 蒸发器
receiver drier 储液干燥器
refrigerant 制冷剂

DC motor 直流电动机
drive mechanism 传动机构
brush 电刷
ventilating system 通风系统
refrigerating system 制冷系统
compressor 压缩机
condenser 冷凝器
expansion valve 膨胀阀
blower 风机

C. Translation

起动机和空调

起动机的作用就是起动发动机，发动机一旦起动，起动机便立即停止工作。起动机一般由直流电动机、传动机构和操纵机构三部分组成。

汽车空调系统主要由通风系统、采暖系统、制冷系统、空气净化装置等组成。其中空调制冷系统主要由压缩机、冷凝器、蒸发器、储液干燥器、风机等组成。

Part Three: Exercises

A. Match the English with Chinese. Draw lines.

1. starter motor　　　　　a. 空调
2. receiver drier　　　　　b. 膨胀阀
3. compressor　　　　　　c. 制冷剂
4. condenser　　　　　　　d. 起动机
5. expansion valve　　　　e. 蒸发器
6. air conditioner　　　　　f. 压缩机
7. refrigerant　　　　　　　g. 储液干燥器
8. evaporator　　　　　　　h. 冷凝器

B. Look at the picture and choose the correct technical terms.

(1)
a. DC motor
b. drive mechanism
c. control mechanism

(2)
a. compressor
b. condenser
c. evaporator
d. receiver-dryer

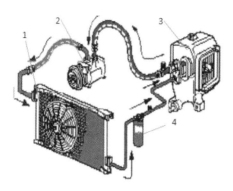

Lesson 3 Illumination, Signal Devices and Instruments

Learning Goals

Grasp the components of illumination, signal devices and instruments.

Part One: Look and Match

Look at the pictures and match them to the terms given below.

headlight   foglight   interior lamp   license plate light   clearance light
tail light   dashboard   speedometer   FG (fuel gauge)   oil pressure gauge
water temperature gauge   horn

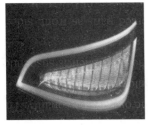
1. _____

2. _____

3. _____

4. _____

5. _____

6. _____

Unit 5 Electrical Equipment

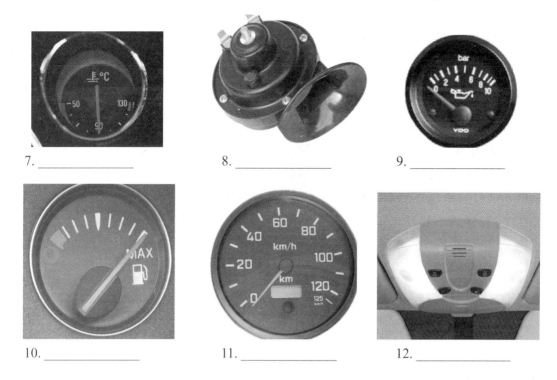

7. _____   8. _____   9. _____

10. _____   11. _____   12. _____

Part Two: Look and Learn

A. Text

Illumination、Signal Devices and Instruments

Modern cars are equipped with various lighting equipments and light signal devices in order to ensure the safety and improve the speed. The instrument is a device that provides the driver with important information about the operation of the car. It is used to indicate the driving of the car and the running of the engine to ensure the normal operation of the vehicle.

lighting equipments mainly include headlights, foglight, interior lamps and licence plate lights, etc. The signal devices generally include the clearance light, taillight, brake light, turning light and backup light, etc. Instruments are generally included oil pressure gauge, water temperature gauge, fuel gauge, speedometer, etc.

B. New words and Phrases

headlight 大灯
turning light 转向灯
foglight 雾灯
license plate light 牌照灯
tail light 尾灯
speedometer 车速里程表
oil pressure gauge 机油压力表
horn 喇叭
signal devices 信号装置

brake light 制动灯
backup light 倒车灯
interior lamp 顶灯
clearance light 示廓灯
dashboard 仪表盘
FG (fuel gauge) 燃油表
water temperature gauge 水温表
illumination 照明
instrument 仪表

C. Translation

照明、信号装置及仪表

为了保证汽车行车安全,提高车辆行驶速度,现代汽车上都装有多种照明设备和灯光信号装置。汽车仪表是为驾驶人提供汽车运行重要信息的装置,用来指示汽车运行与发动机的运转状况,保证车辆正常运行。

照明装置一般包括:前照灯、雾灯、顶灯和牌照灯等;信号装置一般包括:示廓灯、尾灯、制动灯、转向信号灯和倒车灯等;仪表一般包括:机油压力表、水温表、燃油表、车速里程表等。

Part Three: Exercises

A. Match the English with Chinese. Draw lines.

1. horn   a. 大灯
2. dashboard   b. 雾灯
3. headlight   c. 燃油表
4. fuel gauge   d. 转向灯
5. speedometer   e. 喇叭
6. turning light   f. 水温表
7. foglight   g. 仪表盘
8. water temperature gauge   h. 车速里程表

Unit ❺ Electrical Equipment

B. Look at the picture and choose the correct technical terms.

a. ABS indicator

b. charging system indicator

c. brake system indicator

d. transmission warning light

e. door monitor

f. low fuel warning light

g. oil pressure warning light

h. safety belt reminder light

i. airbag warning light

j. check engine warning light

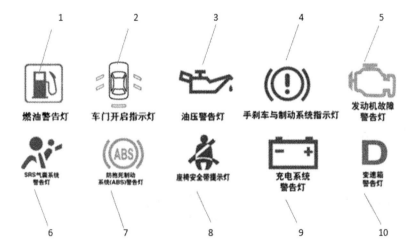

Unit 6　The Car Body

Lesson 1　The Types of Car Body

Learning Goals

Understand the different types of car body.
Grasp the words of components of car body.

Part One: Look and Match

Look at the pictures and match them to the right terms given below.

hard-top convertibles　　2-box sedan　　SUV　　MPV　　hatchback　　station wagon
drophead coupe　　sedan　　sports car

1. _____

2. _____

3. _____

4. _____

5. _____

6. _____

Unit 6 The Car Body

7. _____

8. _____

9. _____

Part Two: Look and Learn

A. Text

The Car Body

The car body mainly protects the driver and compose a good air mechanical environment. A good car body can not only bring better performance, but also reflect the owner's personality. The car body structure is mainly divided into seperate frame construction and integrated body.

1. The car with a seperate frame construction has rigid frame. The body shell is suspended on frame and joined with elastic components.

2. The car with a integrated body has no rigid frame, but reinforced parts as front, side, rear, baseplate,etc. The body and the base frame constitute the rigid space structure of body shell.

B. New words and Phrases

hard-top convertibles 硬顶敞篷车
hatchback 掀背车
drophead coupe 软顶敞篷车
sports car 跑车
integrated body 承载式车身
seperate frame construction 非承载式车身
rigid 坚硬的
suspend 悬挂
reinforced 加固的，加强的
base frame 底架

2-box sedan 两厢车
station wagon 旅行车
sedan 三厢车
frame 车架
mechanical 机械的，力学的
environment 环境
body shell 车身本体
join with 和……结合
baseplate 底板
constitute 构成

C. Translation

汽车车身

汽车车身的作用主要是保护驾驶员以及构成良好的空气力学环境。好的车身不仅能带来更佳的性能，也能体现出车主的个性。汽车车身结构从形式上说，主要分为非承载式和承载式两种。

1. 非承载式车身的汽车有刚性车架。车身本体悬置于车架上，用弹性元件连接。
2. 承载式车身的汽车没有刚性车架，只是加强了车头、侧围、车尾、底板等部位，车身和底架共同组成了车身本体的刚性空间结构。

Part Three: Exercises

A. Match the English with Chinese. Draw lines.

1. hard-top convertibles   a. SUV
2. drophead coupe   b. MPV
3. station wagon   c. 硬顶敞篷车
4. Multi Purpose Vehicle   d. 跑车
5. frame   e. 软顶敞篷车
6. sports car   f. 旅行车
7. Sports Utility Vehicle   g. 车架

B. Look at the picture and choose the correct technical terms.

a. sunroof
b. number plate
c. door handle
d. bonnet
e. side mirror
f. roof
g. trunk lid
h. head light
i. bumper
j. windscreen

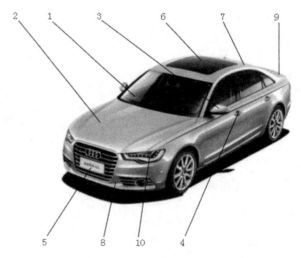

Lesson 2   The Dimension of Car Body

Learning Goals

Understand the different dimensions of cars.
Grasp the words that describe the dimensions of cars.

Part One: Look and Match

Look at the pictures of different sizes of automobiles and match them to the right names given below.

track rear 后轮距   wheel base 轴距   track front 前轮距   overall length 车长
overall width 车宽   overall height 车高   overhang front 前悬   overhang rear 后悬

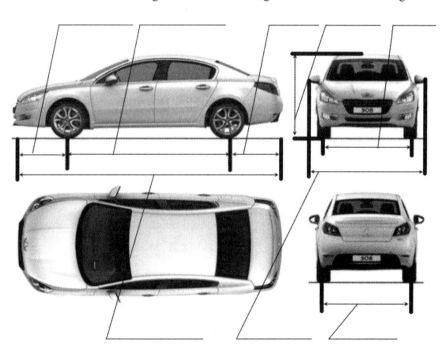

Part Two: Look and Learn

A. Text

Dimensions of Car Body

1. Overall length

Overall length is the most important parameter which influence the use, function and convenience of the car. Therefore, the grade of car body is generally divided by length. The longer the body is, the larger vertical available space the car has.

2. Overall width

Overall width mainly affects the passenger room and flexibility. For passenger cars, if the horizontal layout of the three seats have a wide sense of ride, then the width of the car will generally reach 1.8 meters.

3. Overall height

Overall height directly affects the center of gravity and space. Most cars are under 1.5 meters. In order to create a spacious passenger room, the overall height of SUV is generally more than 1.6 meters.

4. Wheel base

The wheel base is the most important factor affecting the passenger room. In most of the 2-box and 3-box sedans, the seat of the passengers is arranged between the front and rear axles, so longer wheel base makes the passenger's longitudinal space larger.

B. New words and Phrases

track front 前轮距
track rear 后轮距
overall width 车宽
overhang front 前悬
parameter 参数
vertical 垂直的
flexibility 灵活性
layout 布局，排列
spacious 宽敞的

wheel base 轴距
overall length 车长
overall height 车高
overhang rear 后悬
convenience 便利
available 可利用的
horizontal 水平的
gravity 重力
axles 轴

C. Translation

车身尺寸

1. 长度

长度是对汽车的用途、功能、使用方便性等影响最大的参数。因此一般以长度来划分车身等级。车身越长意味着纵向可利用空间越大。

2. 宽度

宽度主要影响乘坐空间和灵活性。对于乘用轿车，如果要求横向布置的三个座位都有宽阔的乘坐感，那么车宽一般都要达到1.8米。

3. 高度

车身高度直接影响重心和空间。大部分轿车高度在1.5米以下。SUV为了营造宽阔的乘坐空间，车身一般在1.6米以上。

4. 轴距

在车长被确定后，轴距是影响乘坐空间最重要的因素。对于大多数的两厢和三厢轿车，乘员的座位都是布置在前后轴之间的，所以长轴距使乘员的纵向空间更大。

Part Three: Exercises

A. Match the English with Chinese. Draw lines.

1. track front   a. 车宽
2. wheel base   b. 后悬
3. track rear   c. 车高
4. overall length   d. 后轮距
5. overall width   e. 前悬
6. overall height   f. 轴距
7. overhang front   g. 前轮距
8. overhang rear   h. 车长

B. Choose the right answer.

1. If there are airbags in a car, _____.

    a. it is enough safety

    b. people needn't fasted the seat belts

    c. seat belts should be fastened

2. Airbag is a_____.

 a. vehicle safety device   b. warning device   c. charging device

3. Airbag is_____.

 a. just a bag which is full of air

 b. a kind of device for comfort

 c. a kind of device for heating

4. The body serves_____.

 a. the obvious purpose of providing shelter

 b. the obvious purpose of providing comfort

 c. a&b

5. The body is generally divided into_____sections.

 a. Two   b. Four   c. three

Part Four: Learn More

Table 6.1　Common Unit of Measurement on a Car

meter (m)	米
centimeter (cm)	厘米
millimeter (mm)	毫米
kilometer (km)	千米
Kilogram (kg)	千克
Liter (L)	升
Watt (W)	瓦
Kilowatt (kw)	千瓦
Volt (V)	伏
kilometer per hour (km/h)	千米/时
revolutions Per Second (rps)	转/秒
revolutions Per Minute (rpm)	转/分
Liters per 100 kilometers (L/100 km)	升/100 公里

Lesson 3　Auto Interiors and Exteriors

Learning Goals

Understand the different automobile interiors and exteriors.
Grasp the terms of interiors and exteriors.

Part One: Look and Match

Look at the pictures of auto exteriors and exteriors and match them to the right names given below.

　　instrument panel　　　seat　　　central armrest box　　　floor board　　　roof　　　ashtray
　　window film　　　sun visor　　　luggage rack　　　foot board　　　S-Line　　　fender
　　steering wheel covers　　　decorative strip　　　glove box

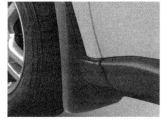

1. _____

2. _____

3. _____

4. _____

5. _____

6. _____

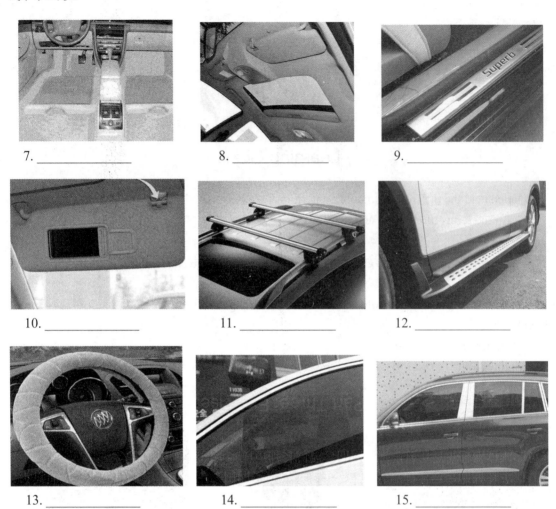

7. _____ 8. _____ 9. _____

10. _____ 11. _____ 12. _____

13. _____ 14. _____ 15. _____

Part Two: Look and Learn

A. Text

Automotive interiors

What we call interiors of a car domestically is the word "Interiors" in English. Since this part of car has a certain decorativeness, so the usual translation within the industry is called "Interiors". But from the English word "Interiors", we can see that these parts are not only decorative, but also the functional, security, and engineering properties are richly involved in.

B. New words and Phrases

interiors 内饰
instrument panel 仪表板
central armrest box 中央扶手盒
roof 车顶
window film 玻璃膜
luggage rack 行李架
S-Line 迎宾条
steering wheel covers 方向盘套
glove box 储物盒

exteriors 外饰
seat 座椅
floor board 地板
ashtray 烟灰缸
sun visor 遮阳板
foot board 脚踏板
fender 挡泥板
decorative strip 装饰条
be involved in 涉及

C. Translation

汽车内饰

我们国内所说的汽车内饰，其实就是英文的 Interiors。由于这一部分汽车零部件具有一定的装饰性，所以业内通常的翻译都叫做"汽车内饰"。但是从英文"Interiors"可以知道，这部分零件不光只有装饰作用，他们所涉及到的功能性、安全性、以及工程属性是非常丰富的。

Part Three: Exercises

A. Match the English with Chinese. Draw lines.

1. interiors
2. exteriors
3. roof
4. floor board
5. instrument panel
6. sun visor
7. luggage rack
8. seat

a. 仪表台
b. 行李架
c. 遮阳板
d. 外饰
e. 座椅
f. 车顶
g. 内饰
h. 地板

B. Look at the picture and choose the correct technical terms.

a. glove box
b. central armrest box

c. steering wheel

d. seat

e. instrument panel

f. shift lever

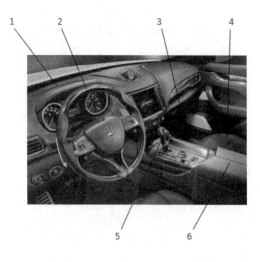

Part Four: Learn More

Dashboard 一词的来历

汽车上的仪表盘，除了叫 instrument panel 之外，还可以称为 dashboard。Instrument panel 一词很好理解，直接翻译出来就是带有仪表的板子。dashboard 一词中，board 是指一块板子，而 dash 的意思则是 "猛冲、泼溅"，这两个意思组合起来，为什么会有仪表盘的意思？

古时候，人们在马车前安置了一块板子用来防止马蹄子溅起来的泥土飞溅到驾驶员身上，这块板子就叫做 dashboard，如图 6.1 所示。

图 6.1 古代马车

早期的汽车由于没有挡风玻璃和顶棚，dashboard 依然保留了下来。汽车再继续发展，一些仪表就被安装到了这块板子上，因此 dashboard 就开始有了仪表盘的含义。确切地从词源上说，dashboard 这个词产生于 1846 年，用于马车，而在 1904 年，延续使用在了汽车上。发展到现在，dashboard 的使用范围也远远超出了汽车仪表盘的范围，还可以用作控制面板、数码仪表板等等。

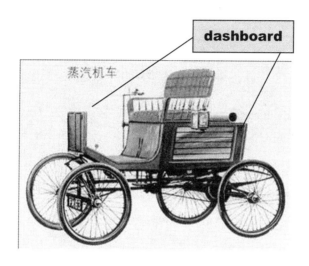

图 6.2　"dashboard" 示意图

Unit 7 Car Maintenance

Lesson 1 Car Maintenance Tools

Learning Goals

Understand common tools in car maintenance.
Grasp the words of common tools.

Part One: Look and Match

Look at the pictures of different types of tools and match them to the right names given below.

ring spanner   combination wrench   torque wrench   adjustable wrench
ratchet handle   slip joint pliers   pair of pliers   needle nose pliers   pipe spanner
snap ring pliers   claw hammer   test pencil   file   rubber hammer   socket
four-wheel aligner   connector   piston ring dismounting pliers   jack   work light
oil filter wrench   extension rod   vise-grip pliers   air compressor

1. _____

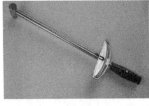
2. _____

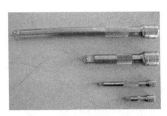
3. _____

4. _____

5. _____

6. _____

Unit **7** Car Maintenance

7. _____

8. _____

9. _____

10. _____

11. _____

12. _____

13. _____

14. _____

15. _____

16. _____

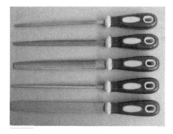
17. _____

18. _____

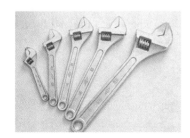
19. _____

20. _____

21. _____

93

22. _____   23. _____   24. _____

Part Two: Look and Learn

A. Text

The Type of Car Maintenance Tools and Tool Selection Principle

The type of car maintenance tools

As a car maintenance technician, the usual job is car maintenance. So it is unavoidable to deal with car maintenance tools. There are various types of car maintenance tools which can be classified as follows:

1. General car maintenance tools

General tools include hammer, screwdriver, pliers, wrench, etc. The commonly used spanners for auto repair are fork spanner, ring spanner, combination spanner, socket wrench, adjustable wrench, torque wrench and special wrench, etc.

2. Special car maintenance tools

Special tools include spark plug socket wrench, oil filter spanner, jack and piston ring dismounting pliers.

Tool selection principle

1. Select special tools first, and the general tools are second choice.

2. The selection principle of wrench: firstly socket wrench and ring spanner, followed by fork spanner, and finally the adjustable spanner.

B. New words and Phrases

tool 工具
combination wrench 两用扳手
adjustable wrench 活动扳手
ratchet handle 棘轮扳手

ring spanner 梅花扳手
torque wrench 扭力扳手
slip joint pliers 鲤鱼钳
test pencil 试电笔

pair of pliers 钢丝钳   needle nose pliers 尖嘴钳
snap ring pliers 卡簧钳   socket 套筒
claw hammer 羊角锤   rubber hammer 橡胶锤
air compressor 空气压缩机   four-wheel aligner 四轮定位仪
jack 千斤顶   vise-grip pliers 大力钳
file 锉刀   connector 转接头
oil filter wrench 机滤扳手   extension rod 加长杆
piston ring dismounting pliers 活塞环拆装钳

C. Translation

汽车维护保养工具类型及选用原则

汽车维护保养工具类型

作为一名汽车维修技术人员，汽车维护保养必定少不了，因此免不了与汽车维修工具打交道。汽车维修工具种类繁多，但是可以总结为以下几类：

1．汽车维护通用类工具

通用工具有锤子、螺丝刀、钳子、扳手等。汽车修理常用的扳手又开口扳手、梅花扳手、两用扳手、套筒扳手、活动扳手、扭力扳手和特种扳手等。

2．汽车维护专用类工具

专用工具有火花塞套筒扳手、机油滤清器扳手、千斤顶和活塞环拆装钳等。

工具的选用原则

1．优先选用专用类工具，其次考虑通用类工具。

2．扳手的选用原则：优先选用套筒扳手、梅花扳手，其次为开口扳手，最后选用活动扳手。

Part Three: Exercises

A. What are we? Please write down the English and Chinese names of the following vehicle pictures.

1._____   2._____   3._____

4. _____   5. _____   6. _____

B. Fill the table with the types of tools by the classification of the following groups.

Types of tools	
general tools	special tools

Part Four: Learn More

Some useful safety signs

Wear a safty helmet　　　　Protect your eyes　　　　Wear protective gloves
戴安全帽　　　　　　　　　保护您的眼睛　　　　　　戴防护手套

图 7.1　Some useful safety signs

Unit **7** Car Maintenance

Wear a dust proof mask　　　Wear protective shoes　　　Use a respirator
戴防尘口罩　　　　　　　　穿防护鞋　　　　　　　　使用防毒面具

图 7.1　Some useful safety signs（续图）

Lesson 2   Car Maintenance Measuring Tools

Learning Goals

Understand the measuring tools in the car maintenance operation.
Grasp the words of common measuring tools.

Part one: Look and Match

Look at the pictures of different types of Measuring tools and match them to the right terms given below.

vernier caliper   outside micrometer   dial indicator   cylinder gauge   feeler gauge
steel rule   multimeter   tyre gauge   cryoscope   cylinder pressure gauge
manifold pressure gauge   pattern depth gauge

1. _____   2. _____   3. _____

4. _____   5. _____   6. _____

Unit 7 Car Maintenance

7. _____

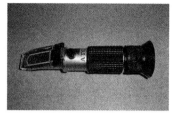
8. _____

9. _____

10. _____

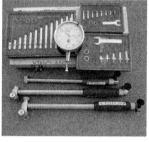
11. _____

12. _____

Part Two: Look and Learn

A. Text

Maintenance Measuring Tools

In automobile maintenance, except for the commonly used measuring rule, calipers and feeler gauge, the main high-precision ones are: vernier calipers, height vernier calipers, depth vernier calipers, etc.

Micrometer measuring tools: micrometer, inside micrometer, depth micrometer, etc.

Meters (indicator gauges): such as dial indicator, micrometer gauge, inner diameter (also cylinder bore gauge), etc.

B. New words and Phrases

vernier caliper 游标卡尺
outside micrometer 外径千分尺
dial indicator 百分表
tyre gauge 轮胎气压表
cylinder bore gauge 量缸表
manifold pressure gauge 歧管压力表

feeler gauge 厚薄规
steel rule 钢尺
multimeter 万用表
cryoscope 冰点仪
cylinder pressure gauge 气缸压力表
rust-proof 防锈

C. Translation

汽车维修量具

在汽车维修中，常用的量具除量尺、卡钳、厚薄规（塞尺）外，精度较高的主要有：如游标卡尺、高度游标卡尺、深度游标卡尺等。

测微量具：如千分尺、内径千分尺、深度千分尺等。

表类量具（指示量具）：如百分表、千分表、内径百分表（亦称量缸表）等。

Part Three: Exercises

A. Match the English with Chinese. Draw lines.

1. dial indicator　　　　　　　a. 外径千分尺
2. bore table　　　　　　　　b. 百分表
3. tyre gauge　　　　　　　　c. 冰点仪
4. pattern depth gauge　　　　d. 量缸表
5. cryoscope　　　　　　　　e. 花纹深度尺
6. multimeter　　　　　　　　f. 轮胎气压计
7. vernier caliper　　　　　　g. 万用表
8. outside micrometer　　　　h. 游标卡尺

B. What are we? Please write down the English and Chinese names of the following vehicle pictures.

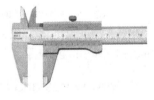

1. _____　　　　2. _____　　　　3. _____

Part Four: Learn More

一、工具量具使用前之准备（Preparation）：

（1）开始量测前，确认工量具是否归零（zero）。

（2）检查工量具量测面有无锈蚀（rust）、磨损（wear）或刮伤（scrape）等。

（3）先清除工件测量面（measuring plane）上的毛边、油污或渣屑等。

（4）用清洁软布或无尘纸擦拭干净。

（5）需要定期检查记录薄，必要时再校正（rectify）一次。

（6）将待使用的工量具及仪器齐排列（arrange）成适当位置，不可重叠放置。

（7）易损的工量具，要放在铺有绒布或软擦试纸的工作台（work bench）上（如：光学平镜等）。

二、工量具使用后的保养（Maintenace）：

（1）使用后，应清洁干净。

（2）将清洁后的工量具涂上防锈油（anti-rust oil），存放于柜内。

（3）拆卸、调整、修改及装配等，应由专门管理人员实施，不可擅自施行。

（4）应定期检查储存工量具的性能是否正常，并作成保养记录（record）。

（5）应定期检验，校验尺寸是否合格，以作为继续使用或淘汰的依据，并将数据作成校验保养记录。

Lesson 3   Complete Maintenance

Learning Goals

Understand the common measuring tools in the complete maintenance operation.
Grasp the related terms of tools.

Part One: Look and Match

Look at the pictures of different types of measuring tools and match them to the right names given below.

headlight   lubricating oil   wiper   air filter   gasoline filter   brake fluid
coolant   brake pads   spark plug

1. _____   2. _____   3. _____

4. _____   5. _____   6. _____

7. _____ 8. _____ 9. _____

Part Two: Look and Learn

A. Text

Automobile Complete Maintenance

As a consumable, the car is to be maintained for every 5,000 kilometers in order to keep its excellent performance. Except for the work of the first class maintenance, the working conditions of the parts of engine and chassis need to be adjusted and checked in the complete maintenance, so as to maintain good technical condition.

B. New words and Phrases

complete maintenance 二级维护
spark plug 火花塞
oil filter 机油滤芯
wiper 雨刮器
consumable 消耗品
first class maintenace 一级维护
check 检查

lubricating oil 润滑油
brake pads 刹车片
air filter 空气滤芯
coolant 冷却液
performance 性能
adjust 调整

C. Translation

汽车二级维护保养内容

汽车作为损耗品，为保持其具有的优良性能，一般每行驶 5,000 公里就要进厂做二级维护保养。二级保养是除执行一级保养作业内容外，调整、检查发动机和底盘各部件工作情况，使其保持良好的技术状态。

Part Three: Exercises

A. Look at the pictures and fill the maintenance items in the box.

the light check   replace brake pads   replace spark plugs   oil change

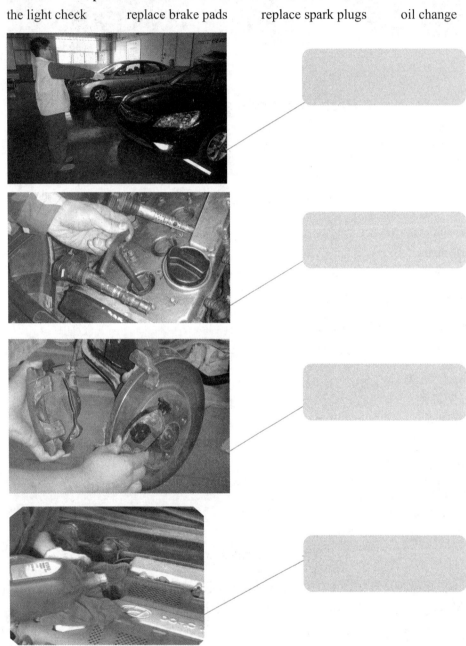

Part Four: Learn More

5S 管理

5S 现场管理法是一种现代企业管理模式。它起源于日本,是指在生产现场(包括车间、办公室)中对人员、机器、材料、方法等生产要素进行有效的管理。目的是通过规范现场、现物,营造一目了然的工作环境,培养员工良好的工作习惯,其最终目的是提升员工的品质。这是日本企业独特的一种管理办法,为日货走向世界立下了汗马功劳。

5S 即整理(SEIRI)、整顿(SEITON)、清扫(SEISO)、清洁(SEIKEETSU)、素养(SHITSUKE),又被称为"五常法则"。

1. 整理(SEIRI)

定义:将工作场所中的不必要的东西尽快处理掉。

正确的价值意识:使用价值,而不是原购买价值。

目的:腾出空间,使空间得到最大化的利用。

注意点:要有决心,及时处理不用物品。

实施要领:每日全面检查;制定"要"和"不要"的判别基准;及时将不要物品科学的清除出工作场所;根据物品使用频率,决定日常用量及放置位置。

2. 整顿(SEITON)

定义:对整理之后留在现场的必要物品分门别类放置,排列整齐并明确数量和有效标识。

目的:整齐的工作环境下使物品一目了然。

注意点:提高工作效率。

实施要领:物品的保管要定点、定量,生产线附近只能放真正需要的物品(不超出所规定的范围放置)。

3. 清扫(SEISO)

定义:将工作场所清扫干净并保持。

目的:使职场明亮,工作人员提高工作效率。

注意点:责任化、制度化。

实施要领:建立清扫责任区;执行例行扫除;调查污染源,予以杜绝或隔离;建立清扫基准,作为规范;开始一次全公司的大清扫,每个地方清洗干净。

4. 清洁(SEIKEETSU)

定义:将办公场所打扫制度化、规范化。

目的:维持职场干净。

实施要领:制订目视管理的基准;制订考评、稽核方法及奖惩制度。

5. 素养（SHITSUKE）

定义：通过晨会等手段，提高员工文明礼貌水准，增强团队意识，养成按规定行事的良好工作习惯。

目的：提升员工的品质，使员工对任何工作都讲究认真。

注意点：长期坚持，才能养成良好的习惯。

实施要领：制订服装、臂章、工作帽等识别标准；制订公司有关规定；制订礼仪守则；教育训练（新进人员强化 5S 教育、实践）；推动各种精神提升活动（晨会，例行打招呼、礼貌运动等）推动各种激励活动，遵守规章制度。

附录　练习题答案

Unit 1　Basic Knowledge of Automobiles

Lesson 1

A. 写出图片表示的汽车种类的英语和汉语名称。

racing car　赛车
limousine　豪华汽车
school bus　校车
tractor　牵引车
MPV　多用途车
tipping vehicle　自卸车
crane car　吊车
minivan　小型货车/面包车
coach　长途汽车
police car　警车
forklift truck　叉车
2-door sedan　双门轿车

B. 根据下表的分类，写出相应种类的汽车种类名称。（参考）

Passenger sedan: convertible, roadster, jeep, SUV, MPV, pickup, coupe, limousine, racing car, etc.
Bus: minibus, midbus, aerobus, coach, double-decker bus, articulated bus, etc.
Truck: van, crane car, tractor, tipping vehicle, forklift truck, platform trailer, etc.
Special-purpose Vehicle: police car, ambulance, fire engine, water sprinkler, sweeper, refrigerator truck, cash truck, oil tank truck, etc.

C. 汽车种类萝卜蹲。

规则：把学生每 5 人分成一组进行游戏，老师将写有汽车种类名称的游戏卡发给学生（每组 5 张），每个人选择一张卡片，将自己的"名字"告诉全班同学。由老师指定其中一人开始，说"XX 蹲，XX 蹲，XX 蹲完，XX 蹲"，被叫到的人继续说，若未说出则被淘汰。坚持最多轮数的小组获胜。每组有 1 分钟练习时间。

Lesson 2

A. 写出图片表示的汽车标志的英语和汉语名称。

Cadillac 卡迪拉克

Peugeot 标致

Porsche 保时捷

Chevrolet 雪弗兰

Fiat 菲亚特

Alfa Romeo 阿尔法·罗密欧

Skoda 斯柯达

Mini 迷你

Chrysler 克莱斯勒

Volvo 沃尔沃

Land Rover 路虎

Jaguar 捷豹

Mazda 马自达

GM（General Motors） 通用

BYD 比亚迪

Mitsubishi 三菱

Lexus 雷克萨斯

Suzuki 铃木

KIA 起亚

Lincoln 林肯

Opel 欧宝

Rolls-Royce 劳斯莱斯

Maserati 玛莎拉蒂

Ferrari 法拉利

Lesson 3

A. 把下列汽车名字翻译成汉语。

1. 君越
2. 林荫大道
3. 昂科拉
4. 朗逸

5. 帕萨特
6. 斯柯达明锐
7. 宝来
8. 思迪
9. 思域
10. 飞度
11. 逍客
12. 索纳塔
13. 世嘉
14. 蒙迪欧
15. 嘉年华

Lesson 4

A. 写出图片表示的汽车种类的英语和汉语名称。

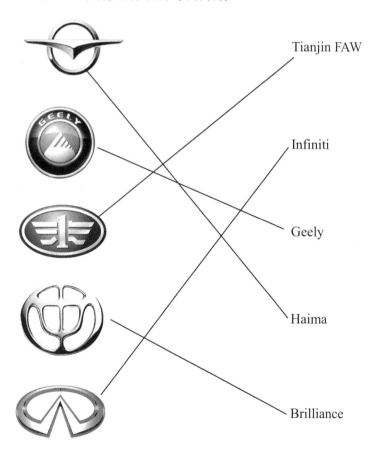

Unit 2　Automobile Sales

Lesson 1

A. 写出 4S 店 4 种功能的英文名称。

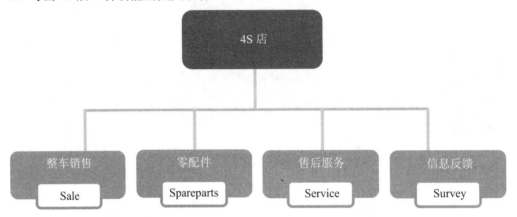

B. 请列举一下当地你所熟悉的 4S 店。（5 个）

以山东省济南市为例：
1. 济南之星（奔驰 4S 店）
2. 济南大友宝马 4S 店
3. 山东金宝利（一汽大众）
4. 济南德辉汽车（进口大众）
5. 润华天信奥迪 4S 店

Lesson 2

A. 通过阅读以下对话，回答问题。

Questions:

1. What is the original price of this model? 这款车原价是多少？
   ¥132,000.
2. What is the price after discussion? 打完折后这款车的价格是多少？
   ¥131,000.
3. Does the customer buy the car? 顾客最后买没买车？
   Yes.

B. 请写出汽车销售流程。

1. 接待（Reception）
2. 需求咨询（Consulting）
3. 车辆介绍（Automobile introduction）
4. 试乘试驾（Test drive）
5. 报价协商（Price discussion）
6. 签约成交（Make a deal）
7. 交车（Auto delivery）
8. 售后跟踪（After-sales service）

Lesson 3

A. 通过上网查询，列举现在的汽车金融公司有哪些？

1. SAIC-GMAC 上汽通用汽车金融公司
2. Volkswagen financial co., LTD 大众汽车金融公司
3. Dongfeng auto financial co., LTD 东风汽车金融公司
4. Ping'an auto finance co. LTD 平安集团汽车金融有限公司
5. Buick financial co., LTD 别克金融有限公司

B. 杨女士最近想购买一辆宝马 320，但是资金不够。通过学习本节知识周后，你认为杨女士可以通过哪几种方式解决资金不足买到自己喜欢的车？

可以通过：

1. 银行车贷 bank loan
2. 宝马汽车金融公司 BMW auto financing co., LTD
3. 利用信用卡分期 Credit Card Installment Payment
4. 申请厂家财务公司的贷款 apply for a loan from Auto Manufacturer finance company
5. 到汽车融资租赁公司办理业务 go to auto financing lease

Unit 3　The Engine

Lesson 1

A. 写出图片表示的发动机种类的英语和汉语名称。

in-line engine/multi-cylinder engine
V-type engine
flat engine
cylinder

supercharged engine
new energy engine

B. 根据下表的分类，写出相应种类的发动机种类名称。（参考）

Fuel: gasoline engine, diesel engine, new energy engine
Stroke: two-stroke engine, four-stroke engine
Cooling method: air-cooled engine, water-cooled engine
Numbers of cylinders: single cylinder engine, multi-cylinder engine
Arrangement of cylinders: single-bank engine, two-bank engine
Method of supercharging: supercharged engine, unsupercharged engine

C. 发动机类型萝卜蹲

规则：把学生每 5 人分成一组进行游戏，老师将写有发动机种类名称的游戏卡发给学生（每组 5 张），每个人选择一张卡片，将自己的"名字"告诉全班同学。由老师指定其中一人开始，说"XX 蹲，XX 蹲，XX 蹲完，XX 蹲"，被叫到的人继续说，若未说出则被淘汰。坚持最多轮数的小组获胜。每组有 1 分钟练习时间。

Lesson 2

A.

1. the fuel-supply system
2. the cooling system
3. the lubricating system
4. the ignition system

B. 用线连接英语与汉语

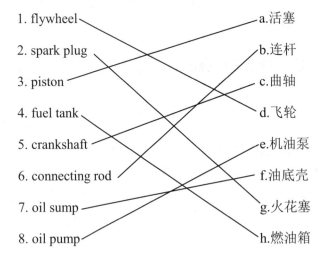

1. flywheel   a.活塞
2. spark plug   b.连杆
3. piston   c.曲轴
4. fuel tank   d.飞轮
5. crankshaft   e.机油泵
6. connecting rod   f.油底壳
7. oil sump   g.火花塞
8. oil pump   h.燃油箱

Lesson 3

A.

1. Intake 进气行程
2. Compression 压缩行程
3. Power 作功行程
4. Exhaust 排气行程

B. 用线连接英语与汉语

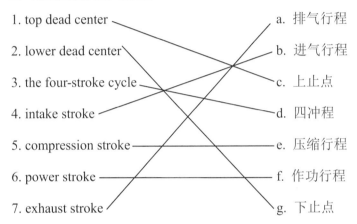

C. 看图写出正确的技术术语

1: spark plug
2: exhaust valve
3: intake valve
4: exhaust manifold
5: intake passage
6: oil atomizer
7: combustor
8: piston

Lesson 4

A. 写出发动机的主要排放污染物

1. CO
2. HC

3. NO$_x$
4. SO$_2$
5. particles

B. 用线连接英语与汉语

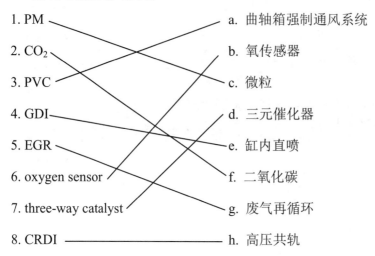

Unit 4　The Chassis

Lesson 1

A. 看图写出正确的技术术语。

1：engine　发动机
2：clutch　离合器
3：manual transmission　手动变速器
4：universal joint　万向节
5：final drive　主减速器
6：drive shaft　传动轴
7：differential　差速器
8：halfshaft　半轴

B. 根据对话选择正确答案。

1. b　　2. a　　3. a　　4. b　　5. a

Lesson 2

A. 阅读课文，完成下列句子。

1. guiding the car to where the driver wants it to go.
2. two
3. assists
4. prevents
5. front

B. 与你的搭档重新排列字母。

manual   suspension   prevent   independent

Lesson 3

A. 朗读课文，完成下列句子。

1. stably
2. external pressure power
3. drum disc
4. service parking
5. large capacity trucks

B. 再次朗读课文，选择正确答案。

1. c   2. b   3. c   4. a   5. b

C. 看图片，选择正确的技术术语。

1. disk brake
2. master cylinder
3. vacuum booster
4. brake pedal
5. drum brake
6. ABS pump
7. brake light

Unit 5　Electrical Equipment

Lesson 1

A. 用线连接英语与汉语

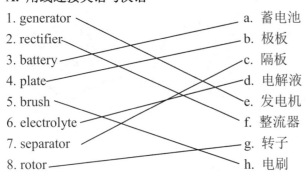

1. generator
2. rectifier
3. battery
4. plate
5. brush
6. electrolyte
7. separator
8. rotor

a. 蓄电池
b. 极板
c. 隔板
d. 电解液
e. 发电机
f. 整流器
g. 转子
h. 电刷

B. 看图写出正确的技术术语

(1)

1: pole 极桩

2: shell 外壳

3: positive plate 正极板

4: separator 隔板

5: negative plate 负极板

(2)

1. front end cover 前端盖

2. brush 电刷

3. rectifier 整流器

4. rotor 转子

5. stator 定子

6. rear end cover 后端盖

7. fan 风扇

8. drive pulley 驱动带轮

附录 练习题答案

Lesson 2

A. 用线连接英语与汉语。

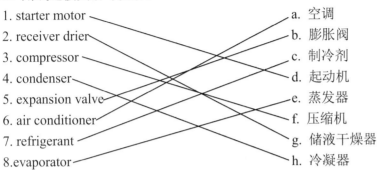

B. 看图写出正确的技术术语。

(1)

1: control mechanism 操纵机构

2: drive mechanism 传动机构

3: DC motor 直流电动机

(2)

1. condenser 冷凝器

2. compressor 压缩机

3. evaporator 蒸发器

4. receiver-dryer 储液干燥器

Lesson 3

A. 用线连接英语与汉语。

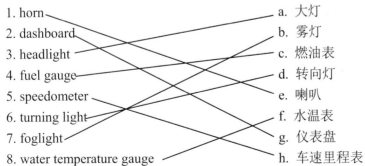

B. 看图写出正确的技术术语。

1：low fuel warning light 燃油警告灯

2：door monitor 车门开启指示灯

3：oil pressure warning light 油压警告灯

4：brake system indicator 制动系统指示灯

5：check engine warning light 发动机故障指示灯

6：airbag warning light 安全气囊警告灯

7：ABS indicator ABS 报警灯

8：safety belt reminder light 安全带指示灯

9：charging system indicator 充电指示灯

10：transmission warning light 变速箱警告灯

Unit 6　The Car Body

Lesson 1

A. 用线连接英语与汉语。

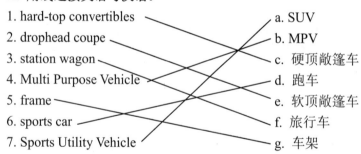

B. 看图写出正确的技术术语。

1. windscreen 挡风玻璃

2. bonnet 引擎盖

3. side mirror 后视镜

4. door handle 车门把手

5. number plate 车号牌

6. sunroof 天窗

7. roof 车顶

8. bumper 保险杠

9. trunk lid 后备箱盖
10. head light 大灯

Lesson 2

A. 用线连接英语与汉语

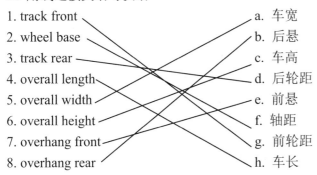

1. track front — d. 后轮距
2. wheel base — f. 轴距
3. track rear — g. 前轮距
4. overall length — h. 车长
5. overall width — a. 车宽
6. overall height — c. 车高
7. overhang front — e. 前悬
8. overhang rear — b. 后悬

B. 选择正确的答案

1. c   2. a   3. b   4. c   5. c

Lesson 3

A. 用线连接英语与汉语

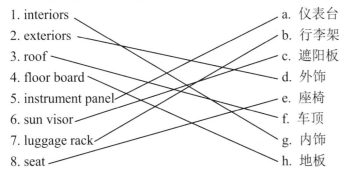

1. interiors — g. 内饰
2. exteriors — d. 外饰
3. roof — f. 车顶
4. floor board — h. 地板
5. instrument panel — a. 仪表台
6. sun visor — c. 遮阳板
7. luggage rack — b. 行李架
8. seat — e. 座椅

B. 看图写出正确的技术术语

1：instrument panel 仪表盘
2：steering wheel 方向盘
3：glove box 储物盒
4：seat 座椅
5：shift lever 变速杆
6：central armrest box 中央扶手盒

Unit 7　Car Maintenance

Lesson 1

A. 写出图片表示的维修工具种类的英语和汉语名称。

jack	千斤顶
ring spanner	梅花扳手
piston ring dismounting pliers	活塞环拆装钳
ratchet handle	棘轮手柄
oil filter wrench	机滤扳手
four-wheel aligner	四轮定位仪

B. 根据下表的分类，列举相应种类的汽车维修工具名称。（参考）

Types of tools

general tools	special tools
ring spanner	spark plug socket wrench
fork wrench	oil filter wrench
torque wrench	jack
hammer	four-wheel aligner
screwdriver	piston ring dismounting pliers

Lesson 2

A. 用线连接汉语与英语。

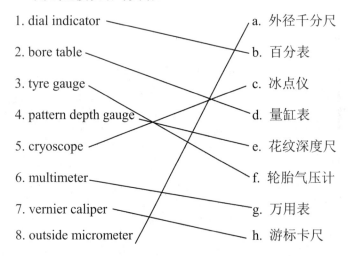

1. dial indicator　　　　　a. 外径千分尺
2. bore table　　　　　　b. 百分表
3. tyre gauge　　　　　　c. 冰点仪
4. pattern depth gauge　　d. 量缸表
5. cryoscope　　　　　　e. 花纹深度尺
6. multimeter　　　　　　f. 轮胎气压计
7. vernier caliper　　　　g. 万用表
8. outside micrometer　　h. 游标卡尺

B. 写出图片表示的维修量具种类的英语和汉语名称。

1. 冰点仪 cryoscope
2. 花纹深度尺 pattern depth gauge
3. 游标卡尺 vernier caliper

Lesson 3

A. 看图，将下列二级维护项目蓝色方框内。

灯光检查　　更换刹车片　　更换火花塞　　更换机油

灯光检查
the light check

更换火花塞
replace spark plugs

更换刹车片
replace brake pads

更换机油
oil change